James Leasor was educated at The
Oriel College, Oxford. In World W
into the Royal Berkshire Regime.
Lincolns in Burma and India, wher   ur three and a
half years. His experiences there stimulated his interest in India,
both past and present, and inspired him to write such books
as *Boarding Party* (filmed as *The Sea Wolves*). He is a former
feature writer and foreign correspondent at the *Daily Express*.
There he wrote *The One that Got Away*, the story of the sole
German POW to escape from Allied hands. As well as non-
fiction, Leasor has written novels, including the Dr Jason Love
series, which have been published in 19 countries. *Passport to
Oblivion* was filmed as *Where the Spies Are* with David Niven.

BY THE SAME AUTHOR
ALL PUBLISHED BY HOUSE OF STRATUS

NON-FICTION

BOARDING PARTY
GREEN BEACH
THE MARINE FROM MANDALAY
THE MILLIONTH CHANCE
THE ONE THAT GOT AWAY
THE PLAGUE AND THE FIRE
RUDOLF HESS: THE UNINVITED ENVOY
SINGAPORE: THE BATTLE THAT CHANGED THE WORLD
THE UNKNOWN WARRIOR
WAR AT THE TOP
WHO KILLED SIR HARRY OAKES?

JAMES LEASOR

Wheels to Fortune

A brief account of the Life and Times of
WILLIAM MORRIS, VISCOUNT NUFFIELD

HOUSE OF
STRATUS

This edition published in 2001 by House of Stratus, an imprint of
Stratus Books Ltd., 21 Beeching Park, Kelly Bray,
Cornwall, PL17 8QS, UK.

www.houseofstratus.com

Typeset, printed and bound by House of Stratus.

A catalogue record for this book is available from the British Library
and the Library of Congress.

ISBN 0-7551-0047-6

For
JOAN

"To any man worth his salt, the desire for personal gain is not his chief reason for working. It is that desire to achieve, to be a success, to make his job something worthy of his mettle and self-respect. Money plays an important part in this – it is stupid to deny it – but it is the part of air to living things…"

Lord Nuffield as W R Morris, 1927

"Inter miracula ipse praecipuum miraculum, inter moventia primum mobile" – *"Among miracles, he is himself the chief miracle; among movement he is the source of motion…"*

Public Orator, of Lord Nuffield as Sir
William Morris, at Oxford, 1931,
when he was receiving the degree of
DCL from the University.

CONTENTS

page

1   The Start of the Matter   1

2   To Cowley, and to War   22

3   The Car you Buy to Keep   37

4   Expansion   51

5   A Car for £100   71

6   Safety Fast!   81

7   £26,000,000 Given Away   90

8   The Man with the Red Wig   103

9   War in Three Cities   115

10 Interim Assessment   127

CHAPTER ONE

The Start of the Matter

I began with nothing. I hadn't even a five-pound note. But although I was only seventeen I felt that W R Morris would pay me a higher salary than anybody else. So I started to work for W R...

Lord Nuffield, quoted in an article, 1926

It was past noon when the hammering ceased.

The young man stood back for a moment, filled his pipe carefully, and lit it, watching the car critically through eyes half closed, as an artist scrutinizes a picture. An unusual car, this, a very special one, with a greenish-grey body, large brass acetylene lamps and a Cape cart hood of scrubbed canvas.

There would never be another car like it, although thousands, even millions of others would bear its name. This one was especially his; he had made it himself.

On this spot in Longwall Street, Oxford, where William Richard Morris planned his first car, garages still stand – a branch of one of his enterprises called, appropriately, The Morris Garages. Some motorists pass by and know nothing about this first car, or what happened before and afterwards. To them, this is just another garage in a narrow, winding road; a place for petrol and nothing more. And if they were told that here a major industry began, and one that changed the face of England, they would probably not believe it.

"Really?" they would say, polite with disbelief. "Fancy that, now." Then they'd pocket their change, and drive off again, as like as not in a direct descendant of that first Morris car, and by the time they reach the main road, most of them have probably forgotten all about the garage and are only intent on passing the car ahead.

But the people who live near by do not all forget so easily. Many remember quieter days, when a car in the streets of Oxford, lifting the dust and making the horses shy in the shafts, was something to talk about.

They remember young William Morris working there, sleeves rolled up, first repairing motor cycles, and then building his own; and then repairing cars and finally building those, too. The site of the garages has long been connected with wheeled vehicles. Before the coming of cars, coaches were kept in sheds there, and Morris used some of these sheds and stables as workshops, and the rest as garages which he let out to local motorists. In place of the mild, friendly sound of horses other, stranger, noises arrived: a sudden cacophony of sound as an engine started, or the sharp, unexpected bang of a backfire that would frighten the tame deer in the grounds of Magdalen College across the road.

These were new noises that had come to stay; and new smells ousted the mellow tang of leather harness: the smell of petrol and oil, and blue exhaust smoke.

From this small start an organization has grown and multiplied and advanced, so that today, instead of a handful of men working together on an idea in which they all believed, there are thousands; and instead of a garage off the High Street, with a tree growing up in the yard and sticking its branches through the upper windows, the factories producing the cars that bear the founder's name, and other makes besides, cover more than one million square yards.

The enterprise and perseverance of that one man has changed the face of Oxford and transformed it from a quiet University city to one of the country's most important industrial areas, yet he still smokes cigarettes he has rolled himself, and has not let success change him very much.

His title is now Lord Nuffield but he remains Will Morris in outlook.

Nowadays, he drives an eight horsepower Wolseley instead of riding a bicycle, but the change is only from two wheels to four, and, anyway, his factory made the car just as he had made the cycle…

His hair is nearly white, like silver wire, but brushed straight back from his forehead as it always was. His eyes are pale blue with large, dark intense pupils; he is not a man people can easily look at and lie to. Physically, he is no more impressive than he was as a lad, but he is as tough as a steel spring.

On the mantelpiece in his office at Cowley is a model ship's wheel, won within the last few months at deck tennis on an ocean liner. This success pleases him far more than talk of all the money he has made – although he has given away more than £26,000,000. He has only a very small fraction of this amount left to his name now, but he is not worried; he goes on giving, for his philosophy is simple and entirely genuine.

"What does a great deal of money mean to anyone, anyway? It's only a worry," he says, and he speaks from his own experience; the millions he made all weighed heavily on him.

"I started, earning five shillings a week," he is fond of saying. "I could go back to the beginning again tomorrow – and I might be happier if I did…"

He did not set out to make money, and he has often been surprised at his own success. It has brought many problems, most of them undreamed of when he made that first Morris car. For instance, he is continually being pestered and worried by the writers of begging letters; by people who want him to finance schemes of all kinds, from extracting gold from sea-water to mounting expeditions to search for the treasure lost in sunken Spanish galleons, and by many others all after his money.

"Why should you keep on worrying?" he asks other men of means. "The best thing you can do with money is to give it away. Rich men don't give nearly enough money away…"

"Well, why *keep* it?" Lord Nuffield persists, for it is a favourite theme of his. "What can you *do* with it? You can only wear one suit at a time. You can only eat one meal at a time. You lose even the pleasure of wishing for things.

"If you have so much money that you can buy anything you want, you find you don't want anything.

"Nothing gives you any satisfaction unless you have to struggle a bit to get it, and then, when you've got it, you don't want it…

"Look at me. People say to me, 'What can I give you for your birthday?' meaning the question in a kindly way.

"But what can I reply that is truthful? There's nothing I want. If I wanted anything, then obviously I would have bought it. I can *afford* to.

"I have fewer clothes in my wardrobe than most men. Some of my suits are ten years old.

"There's nothing at all that I want – anywhere."

This is one measure of the triumph that a life of hard work and calculated risks, larded with a genius for making a right decision, has brought to William Richard Morris: the position of being able to afford anything he wants, only to discover that nothing he still desires can be bought with money.

In the last twenty years, honours and awards have been heaped upon him, but it is doubtful whether he really wanted any of them, for he is still, and always has been, a simple man of quiet tastes: the questing mechanic in search of the perfect machine.

Nevertheless, the honours have arrived, unasked and unsought. Before the war, he was made a Fellow of the Royal Society; and, after it, a Fellow of the Royal College of Surgeons. He is an honorary Master of Arts and a Doctor of Civil Law of Oxford University; an honorary Doctor of Law of the Universities of Sydney and Birmingham, of Melbourne and Belfast; a Doctor of Science of New South Wales University of Technology; and an honorary Freeman of five cities.

Lord Nuffield is also an honorary member of the British Medical Association, of the Worshipful Society of Apothecaries, of the British Orthopaedic Association, and the Faculty of Anaesthetists. He is an honorary Fellow of Pembroke and Worcester Colleges; President of Guy's Hospital, a life governor or president of hospitals in other parts of the country, and the holder of many more titles than these. Yet he has never attended any university as a student, and he received all his schooling at the village school in Cowley, within shouting distance of the great factory where his cars are now made.

"The only road to success is hard work and, of course, foresight," he says now, looking back on his life. "It's not always the men who've had an expensive education who do things," he will add, almost defensively; and indeed he is one man whose whole life has proved this point.

He was born in Worcester on 10 October 1877, to Emily, wife of Frederick Morris, and christened William Richard a few days later. That year work actually started on a tunnel under the channel to link England with France, a project that has still not been completed. W G Grace was at his peak; Blondin, "The Hero of Niagara", was appearing nightly on a high wire to audiences of thousands at the Crystal Palace, and what was called "a serious, not to say alarming, state of insecurity" existed in Clapham, where gangs of burglars were using builders' ladders to climb up to the bedrooms of the big houses. At St Austell, in Cornwall, experiments were being made with a telephone in a tin mine which, said one expert, "showed that it may do better service than in being the medium for the conveyance of songs and jokes from a distance".

Looking back now, this prophecy seems a very easy one to have made, but at the time there were not many who believed that it would eventually be possible to speak to people in distant towns and countries. There would have been far, far fewer who could ever think that the son and heir of Frederick Morris would become a man famous and wealthy, or – had anyone prophesied this – that his name would be driven round the world.

As with nearly all men who have achieved outstanding success, little is known about the early life of Lord Nuffield. When he was a boy, there was no reason to suppose that his life would be much different from that of any of the neighbours' children. When it became vastly different, then people were quick to recall how they had "always thought he would do well". Had Morris failed wretchedly in his first venture on his own, these same people, of course, would have been quick to call witness to his folly.

One of the most common myths about the background of Lord Nuffield is that he was born and reared in poverty. This is quite untrue; both his parents were educated middle-class people. His father, Frederick Morris, was an accountant; his mother, the daughter of an Oxfordshire farmer.

The Morris line is a long one, and can be traced back nearly seven hundred years. The first William Morice held land in Swarford "of the manor of Hook Norton" in 1278. Three generations later the name was written as it is today, and with each generation there was a son called Richard or one called William; but never one to equal the last of the line, who bore both these Christian names.

The family were tenants, not land-owners; but middle-class farmers, who were churchwardens in their local churches. All of them, on both sides of the family, were born and died within the county boundaries of Oxfordshire.

Frederick Morris was an adventurous man, although he has suffered, as do all fathers of the famous, in being remembered not so much for what he did himself, but for what his son has done. As a young man he went to America, and later drove a mail coach across the prairies in Canada before the days of the railroad. He had travelled widely, and was actually made a "blood-brother" in a Red Indian tribe. When he was twenty-seven, he came home to marry Miss Emily Ann Pether, of Wood Farm, Headington.

She had been born in Cowley, in those days a small village only a few fields away from Headington, and this is the first time that Cowley appears in connection with the Morris family. Later, of course, Morris and Cowley were to be associated in the name of Morris-Cowley, one of the most popular British cars of all time; but this was not yet. There were no blast furnaces in Cowley then, no assembly plants or paint shops: there were instead, a smithy and an inn, a church, a vicarage and a handful of houses. There were no cars along the Cowley Road, either, for in those days there were no cars anywhere, only ideas in the minds of a few dreamers that some day, somehow, carriages would move without horses.

When William Morris was still a very small boy of three or four, his family moved from Worcester back to Headington, and there he grew up. Headington has changed from a village of character into a straggling and rather undistinguished suburb of small semi-detached houses outside Oxford on the London road. Many of the people who live there work in one or other of the Nuffield factories, and the town grew without much planning in the twenties and the thirties as the Morris works spread. When Morris was a boy, though, most of the people in Headington were farming folk, and young William would walk across fields, by footpaths, when he went to school at Cowley.

Many people who have discovered where he was educated have tried hard to turn this knowledge to their own advantage.

"You'll remember that I was in your class at school in the old days," they write, but Lord Nuffield remembers no such thing. "If all those who claim they went to school with me at the church school were put together they would need the largest classroom ever built," he says. "There were only six in my form."

He has a long memory for his boyhood, for he is a great sentimentalist. His office, indeed, was once the parlour of the headmaster of Hurst's Grammar School, where his father had been a pupil, and this old school is still a part of the Morris buildings. The chapel remains, and so do the dormitories, and when the works were being constructed he gave orders that as much as possible of the original stonework should be left undisturbed.

The old headmaster would approve of Lord Nuffield's office, could he return to see it. As the office of the Honorary President, which is his title nowadays, it is in the tradition of scholarship rather than big business: a quiet room for a quiet man – a desk, a fireplace, a sofa and some chairs. On one wall, above his desk, is a large picture of Magdalen tower seen from the bridge. This is the first view Lord Nuffield had of Oxford, coming in from Cowley to start a business and to make his name.

Once, at a meeting of motor dealers, he surprised them by telling a story of his childhood. On his way to school, he had to walk round three sides of a private cricket field, and one day he climbed the railings that fenced this, and scampered across the grass.

"When I felt my feet upon that springy turf," he said, "I swore that one day I would own that field. Gentlemen – I do…"

Of course he does, for he is one of the few men of this century who has accomplished all the material things he set out to do when he was a young man, and much more besides.

He left school when he was sixteen, and because his parents were anxious for him to learn a trade, he was apprenticed to a cycle-maker and repairer near his home. Years later, when he gave £1,380,000 to Oxford University for medical research, he recalled that his own first ambition was to have been a surgeon.

"But there was no money to be a surgeon," he said simply, "so I had to take on the next best thing – mechanics. One thing I wanted to do was to use my hands…"

Many of those who grew up with him do not remember this early ambition. There was no hope of his achieving it, and so he did not speak about it outside his immediate family circle, but nursed it himself, a private, secret wish; one of the very few he would never be able to gratify, for by the time he had made enough money to pay for medical training, the impetus of his enterprise had carried him beyond the possibility of being a student. He fitted reverse gears to all his cars, but he could not go back a step himself.

Of all his contemporaries, though, William Morris probably used his hands to the best advantage. First, he used them to make things, and the things he made, in their turn, made money, and then with the money he helped others.

When he was a lad, cycling was the great craze of the day, rather like television is now; and, in 1893, cycle-making was considered a progressive business. Young Morris, who was being paid five shillings a week for his services, did not consider that it was progressive enough for him, however, and within nine months he had stopped working for others and was working for himself.

"Can I have a shilling a week rise?" he asked his employer. "No," replied that worthy, "I'm paying you what you're worth." Morris did not agree.

He had saved up about £4 in these months of apprenticeship, and with this capital he began in business on his own.

"I was with the bicycle-maker for nine months," he recalls, "and that was time enough to teach me that things were not difficult to do, if only one were *determined* to succeed."

This fierce determination stayed with him all his life, and it recalls the same driving force that consumed Cecil Rhodes.

Someone once asked Rhodes, as a young man, for his motto.

"Do or die," he replied promptly, and Morris felt the same way about life.

"If I go so far as to say 'Do it', and have the authority to say so, a thing *must* be done," he has always said. "Given the will, one can do almost anything.

"Sometimes people come to me and say they cannot do this or that. I tell them I have no use for those who cannot do something they are expected to do."

Such faint-hearts have never been welcome in any of his establishments, for he drives himself hard, spurred by an inner impetus, and he has always expected others in the team to be similarly inspired. If they are not, then they can go elsewhere.

"I work my full weight, and I expect my men to do the same," is how he puts it. Yet, for all the hard streak in his character, which he has needed time and again in near-defeat, he is by no means so hard a man as he has been made out, or as he likes to picture himself.

Lord Nuffield has the reputation of being ruthless with people who make mistakes, but this is not always so. He is much more ruthless with people who disagree with him, for although he can forgive mistakes, differences of opinion he finds it harder to accept. He feels that a man who has never made a mistake will probably never succeed, for everyone at some time or other has made mistakes that probably could have been avoided.

"I go on the theory," he told the Marquis of Donegall during the war, "that if I've got a good man, and the bloomer he makes costs me half-a-million pounds, it would be silly to sack him.

"I will probably get someone not as good who would repeat it. It is unlikely that the first chap will be fool enough to lose *another* half-million pounds in exactly the same way..."

It was probably his distaste for opposing opinion that led him to confess, at the same time, "I've never been able to work with a partner, because I've always found that first impressions are best, whereas partners so often do not agree and dissuade you from them."

As early as 1893 he made this discovery, and, as a result, he decided to go it alone. This was a big decision to take. A generation still separated him from the time in 1935 when he would offer shares in Morris Motors to his employees, with the words, "I was becoming tired of working for myself...now it is my great pleasure to work for others..."

This was one of the two major decisions in his life, although the second was probably the greater. In the slump of 1921, he drastically cut the price of his cars, because he could not sell them, and within weeks was selling thousands, proving again the lesson that Henry Ford had already learned with cars, Lipton with tea, and Woolworth with many things: cut your price and watch your business grow.

But all this was still ahead of him, and when he decided to work for himself, as many other lads have done before and since without notable success, he was taking the first step out of the ranks; the hired hand was no longer up for hire.

Morris was not yet seventeen, he had only served a part of his apprenticeship, and he had no one save his parents to guide him. Like Pitt, he was guilty of the crime of being a young man, and in the Oxford of the late nineties there was a far bigger gulf between Town and Gown that there has ever been since. He was a tradesman, and a newcomer at that. It was a worrying time for him. Probably there were many evenings when he felt so low in spirit that he could have been talked out of the whole undertaking, but he kept on. He always was a sticker.

His first workshop was a slate-roofed, brick lean-to at his home, 16 James Street, Oxford, where in the words of his advertisements all cycle repairs were "promptly attended to". So prompt, indeed, was his attention, that he had to work his own hours, which meant that he started early and finished late every evening. No matter. He was working for himself; he was content.

In Oxford, then as now, bicycles were very popular, for the countryside is flat and the going easy. W R M's repairs were good, and so was his workmanship, and cyclists recommended him to their friends: slowly, but steadily, his reputation grew, and with it, his business.

There was a slogan at the time in Oxford:

Hire a bike from Eyles and Eyles,
And ride like – for miles and miles.

Eyles and Eyles have prospered, too, and now own one of the largest garages in the area; but that is another success story altogether.

11

The demand for their bicycles made Morris realize that he could do more business if he also had some cycles which he could hire out, and so he began by charging sixpence an hour, or five shillings a day for them.

A Scots hairdresser, Mr Austin Medcraft, who was in business in Oxford at this time, remembered for the rest of his life one thing above all else about Morris in these early days. "He was a *terrible* worker," he would say years afterwards. In Scotland, the word "terrible" has not always the same meaning as it has in England: here it meant "terribly hard".

In the evenings, Mrs Morris would come down the small garden to see her son. She was very proud of his independence, and he never forgot the encouragement that both his mother and father gave him in those early days when the neighbours, content to have their sons work for other men, thought he was foolish to start out on his own, and prophesied an early failure.

"I am always grateful for having had parents who encouraged me to carve out my own career. They didn't help me financially, but when my mother saw that I had a natural aptitude for engineering, she encouraged me," he says. "She used to come to my repair shop when I was only seventeen. She wanted to see for herself how I was getting on..."

In those days, manufacturers would supply wheels and frames for cycling enthusiasts, so that they could make their own machines instead of buying them ready-made and, incidentally, save money by so doing.

William Morris decided to build bicycles, too, as well as repairing and hiring them. He calculated that, with low overheads, and working his own hours, he could probably produce them at a competitive price, and he was quite right. There was no electric light at that time, but he had gas which he used for brazing the frame tubes of the cycles, and at night he worked on by its light. He didn't mind how long he worked: he was in business and determined to make it pay.

Soon the shed became too cramped for him, and he moved into Oxford, nearer his customers, renting a shop at 48 High Street, near the Queen's College and opposite the Examination Schools. Outside, he had a sign painted: "William Morris, Cycle-Maker and Repairer."

The move into Oxford was oddly symbolic. He was among all the colleges, within the University, but not of it; a position he has always maintained. Always he has been "Town" and not "Gown", and when he came to choose a badge for his car it was the crest of Oxford that he took – but the City, not the University.

One of his earliest customers was F E Smith, then an undergraduate at Wadham, and later Lord Birkenhead. When Morris was made a peer himself he was disappointed that he could not call himself Lord Morris: that title was already held by a Canadian peer.

"I'm sorry I'll lose my name," he said, "and I don't want to choose a new lengthy title, but anyway, I hope my friends will still call me 'W R' in the same way that Lord Birkenhead's friends always spoke of him as 'F E.'" They did. Always to his intimates he has been "W R" or "W R M"; to others, Will or William; but never Bill or Billy.

Looking back on that first move, now, he recalls: "I built push-bikes and sold them to my acquaintances. In the first year, I turned out and sold fifty – and good cycles they were, too."

It was not a very rewarding job financially. Lord Nuffield calls it "a bare living" nowadays, but he was young then, and single, and ambitious. Nothing was too hard for him.

"I felt I could succeed as my own boss," he says, and so he did, splendidly and magnificently, although success did not come easily, or early. This is something that is often forgotten when Lord Nuffield's career is discussed. People assume that he became rich when he was still a young man, which is not so at all. He was not fully out of the financial wood of debts and overdrafts and worries until he was forty-five. His intense faith in himself, even though others called him pig-headed, carried him through nearly twenty years before his ideas were all vindicated and he became rich.

The first customer Morris had for one of his cycles was the Rev. Francis Pilcher, Rector of St Clement's for thirty-five years from 1878. He bought a bicycle for "about £10".

Mr Pilcher was 6 feet 3 inches tall. "He must have been the tallest man in Oxford," Lord Nuffield recalls. "In those days we measured people for their cycles. His machine had a 27-inch frame and 28-inch wheels – the largest cycle I ever built for a customer."

Years later, Lord Nuffield met Mr Pilcher's son, then Bishop of Sydney, on one of his trips abroad. He told him that he had bought back his father's cycle.

"A clergyman came to see me one day, and said that one of my machines was being sold in a rummage sale, so I bought it back for £10," he explains. "It was the first one I'd ever made."

He still has it in his offices at Cowley; a tall, black-painted machine with caliper brakes.

"You could ride it now," Lord Nuffield tells visitors when he shows it to them, "except that the tyres are perished..."

The remark recalls a similar saying of Henry Ford, who, when he became a millionaire, still kept his first tiny Ford car under the bench of his own private workshop as a kind of token, and a reminder of how it all began.

"It would still work now, if they hadn't taken bits off for souvenirs," he would say when people prevailed on him to show it to them.

Shortly after W R M set up in business in Oxford, the local postmaster, Mr Enos Hughes, agreed that he should have the contract to repair the cycles used by the local telegraph boys.

"I admired his good work and affability," he explained afterwards, when someone asked why such a newcomer should have this in preference to firms that had been established longer.

Lord Nuffield loathes personal publicity of any kind, but he has always had an astonishing flair for advertising his product;

whether it was equipping the then Prince of Wales on his shooting trip to Africa with Wolseleys in the early thirties, or backing MG's in their record-breaking racing efforts later on.

When he was building cycles, then, he built two kinds: one to sell and one to race. He knew that if his cycles won races they would be discussed and become popular with enthusiasts. This fact was remembered years later when he marketed the MG sports cars. They won race after race until their reputation for speed and reliability made them the most sought-after small sports car of the day – although, oddly, he never had the same interest in car racing that he used to have for racing cycles, and was inclined to listen indulgently to those who would make MG a name famous on the race tracks of the world.

As a cycle maker, Morris rode his own machines to championship meetings in the county, and also in Buckinghamshire and Berkshire. Most cyclists are good for one distance only, for long or short races; Morris was astonishing in that he was champion at *all* the distances, from half a mile to fifty miles. His energy was tremendous. All day he would work, from seven in the morning until any hour at night. At the weekends he would cycle to the race, and then proceed to win it.

In 1900 he held all seven speed championships for cycle racing in the three counties of Oxfordshire, Buckinghamshire and Berkshire.

He has always kept this first enthusiasm for cycling, and when he opened the Bicycle and Motor Cycle Show at Olympia in 1936, he astonished those present by taking a cycle from one of the stands, borrowing a pair of trouser clips, and riding off round the hall.

Nowadays the medallions he won in his teens and twenties hang in a special frame in his office. There are many of them: each one engraved with a date and a place, and each a testimony to his immense drive and enthusiasm, and also to his courage and endurance.

In 1902, when cycle racing was threatened by commercialism, Morris stopped entering for the races. After all, he had proved his point: Morris cycles were as good as the best. QED.

By this time he was selling cycles as fast as he could make them; but cycles were already becoming too slow for him. He was way ahead, living in the future, impatient with the present. Motorcycles were just becoming popular, and although they were unreliable, with slipping belts, cycle-type brakes and spluttering, tiny engines, which on hills demanded what their makers called "LPA" – light pedal assistance – they were the advance guard of a whole age of motoring.

"Nineteen-hundred brought a big step forward for me," he recalls. "I designed, built and sold the first Morris motorcycle. I had saved up £2,000 from my earnings on push-bikes. It seemed a time to go forward..."

A few private motor cars were on the roads, for the Act that made it compulsory for a man carrying a red flag to walk in front of every mechanically-propelled vehicle had been repealed in 1896, but the cost of the cars kept them out of the reach of all but wealthy owners, who could afford to have chauffeurs to maintain them.

Motorcycles, on the other hand, were relatively cheap to buy and run, and so consequently more people used them. Morris felt that they would become even more popular with the years, and so he turned his attention to them.

Morris motorcycles, like most other contemporary machines, were of very simple design; basically just a strengthened cycle frame with stronger wheels to take the jerky thrust of the engine. He built the engine himself because, as he explained afterwards, by making it he felt he would learn everything there was to know about engine construction. At this time he had never received any engineering training, nor had he read any complete textbook on the subject. He simply bought some castings and bored out the cylinder on a small treadle-lathe – an extraordinary achievement.

16

"I'm just a born engineer, and I can't help it," he explains to people who ask him how he was able to do this. "Once I see a thing made, then I can make it, but I dislike learning from books. I have to *do* the job and learn from experience."

He had a new sign painted: "Morris Cycle Works – Cycles and Motors Repaired." The business prospered.

Years later, in 1931, Dr Norman Joy, an ex-medical officer of Bradfield College, Berks, told a public meeting that in 1902 he was probably the first doctor in England to use a motorcycle regularly, and when it broke down he could find no one near at hand who was able to put it right. He took it in to Oxford, and there he was told that if anyone could repair it, then that man was William Morris – Sir William by 1931. W R M repaired the machine immediately, and the doctor rode it home.

He found Morris a slim, wiry man, his hair brushed straight back from his forehead without any parting, his eyes light blue, with unusually dark and intense pupils. He had a wry, almost quizzical expression, and talked in staccato bursts, throwing the words out of the side of his mouth as though he were too intent on the job in hand to have time to spare for speech. Morris liked *doing* things; talking he could safely leave to others.

He entered one of his motorcycles, with chain drive and a three-speed gearbox, for the 1904 motorcycle show, which was being held in the Agricultural Hall at Islington. Something went wrong at the last moment, and it was thought that he would not be able to complete the machine in time for the show. He knew differently. For four days and nights he worked without any proper sleep, and when the show opened his motorcycle was there on the stand. Mr Morris very nearly was not there with it, however. He was so tired when he reached London that as soon as he sat down in the warmth of the Underground, he fell asleep, and went round and round the Inner Circle line until a friendly porter woke him up.

This year was also important for another reason. It was the year of his marriage to Miss Elizabeth Anstey, an Oxford girl and the daughter of a local businessman.

No one in public life and affairs has kept more in the background. His wife is rarely photographed, and only sometimes does she go out shopping in Oxford, or else slips into the local repertory theatre unrecognized and leaves before the lights go up. She has travelled widely with her husband, but they do little entertaining, and they have never lived on a high level in the past.

Once a year the Viscountess Nuffield leaves her home on the edge of the golf links at Huntercombe and drives to Cowley to present the prizes at the Morris Motors sports day. Apart from this, she is active in church work, and likes her garden, and seeks no life other than the quiet, reserved one she now has. Sometimes she accompanies her husband, when they drive in a car of his own make to receive some new honour that is being paid him. No one save Lord Nuffield can say how much he owes her, and he has not spoken yet.

After the motorcycle show in 1904 orders began to come in quickly, but although he went on making motorcycles until 1910, he was already tiring of two wheels and looking towards four: he wanted to make motor cars.

Most of the cars then on the roads were foreign; French or German or Italian, and all were so expensive to buy and run that just to own one was a mark of wealth. Morris had been over to the Continent several times to collect cars for buyers in Britain; he had examined them carefully, and in his opinion there was much room for improvement.

Many people still regarded motor cars as something of a joke, a passing amusement, as the velocipedes had been a hundred years before. One of the most popular music-hall songs of the day dealt with the trials of motoring. It was "Get out and get under..."

Others were more openly hostile: Lord Queensberry asked the House of Lords for permission to carry a revolver, "to defend myself against motorists".

The late Lord Montagu of Beaulieu, who in 1906 was editing a motor magazine, *Car Illustrated*, produced a small book on the art of good driving, and in it he gave motorists this advice: "Everything may be comprehended in the phrase – that a motorist cannot go far wrong if he drives like a gentleman; in other words, with consideration for all..."

Nevertheless, motorists were violently unpopular. Cars trundled along the narrow, unmetalled carriage ways, alarming pedestrians and covering them with dust. It was not until the Road Board was set up in 1909, and money was spent on tarring the roads, that the nuisance of flying clouds of dust was conquered. By then, though, cars had come to be associated with selfishness and wealth, and feeling against motorists remained high for some time – and in some places and among some people has still not altogether died away.

As in every other field of human endeavour and enterprise in this country, there were plenty of people who said that the car was not an emblem of "real" progress, but only of progress towards disaster. Looking back now, and being wise after the event, it seems incredible that so few people believed that the motor car had a future. But, as with wireless and flying in their early days, the general feeling was that it would all come to very little. We had used the horse for hundreds of years; nothing would ever replace it on the roads. Only a handful of men thought differently, and they thought on their own. One of these men was William Morris.

On his journeys round the countryside, first by cycle and later on his home-made motorcycle, he had seen many of these early cars chugging along, high up off the ground, their occupants goggled and veiled against the dust which blew up from the unmade roads. He knew that the market was small, of course, at the time, for only a few could afford cars, but he believed that if a car could be produced at a reasonable figure, then it would not just be a plaything of the rich, but a means of transport for almost everyone.

If this could be done, then the potential market would be immense. The more he thought about this, the more the idea appealed to him.

The first use of cars for purposes other than pure pleasure was made by doctors, some of whom began to drive to see their patients instead of walking, riding or going by horse and trap. This custom did not become at all general until the financial status of the doctor was raised by the passing of the National Insurance Act of 1911.

To many people it seemed that the words of the prophet Nahum were coming true: "The chariots shall rage in the streets. They shall justle one against another in the broad way; they shall seem like torches, they shall run like lightning..."

By the time he was turning his attention to cars, Morris had made his third move, and was working in the set of stables in Longwall Street, where there is now a branch of The Morris Garages Ltd. Local motorists knew him for a good mechanic, and brought their ailing cars to him. Often there were half a dozen parked in the yard at a time, so he had plenty of opportunity for studying the different makes, and the more he saw of them, the more he felt he could build as good a car as any of them.

The idea took root and grew into action; in 1910 he started work on his first car.

He discussed his intention with a man called W H M Burgess, a wholesale agent for White and Poppe, a Coventry engineering firm. They both agreed that there was going to be a great demand for any car that could be sold cheaply, but to cut the cost it would have to be made in great numbers.

Morris had not the money to order the parts he needed in any quantity, so Burgess promised to sound out as many possible agents as he could as he toured the country for his firm. He succeeded in persuading a number of them to pay deposits towards the cars they would take. He made a note of their names, and the amounts of their deposits in a penny exercise book which is still in his son's possession.

Morris had £4,000 saved up when he started to build his cars, but there were many problems still to overcome. How powerful should the engine be? What was the safest width in proportion to length? Was the chassis strong enough not to "whip" on a rough road?

Morris had little use for books and theories and calculations; he had to try, and fail, and dismantle and try again, and go on trying until he found the right answers himself.

The first car took a long time to grow. He spent two years planning, testing, perfecting and then starting all over again. He would be satisfied with nothing less than the best he could give. He wanted not just another car, but *the* car; a car to which he would be proud to give his name. He knew how he wanted it to look and perform. He wanted a car which the owner could look after himself not one that was complicated and high up off the ground, but one of simplicity; low-built and graceful, not a Ford with an Oxford accent, but a car with a character of its own.

At last it was finished; a grey-green body with a curved brass radiator, thick-spoked artillery wheels, the bright parts gleaming with polish. The stables in Longwall Street smelt of varnish and new leather: an exciting aroma, the smell of a new car, of a whole new world.

On that morning when the car was finished, the last tack hammered into the upholstery, the dashboard shining, and nothing more to do, he stood looking at his handiwork, savouring it.

He has always found it hard to conceal his emotions and so, proud of his work, he ran out into the street and buttonholed the first man he saw.

"Come and see what I've got," he said proudly.

"What?" asked the passerby, unimpressed.

"*A car*," replied Morris, grinning. "And – I've made it myself."

The other man snorted his disbelief.

But the fact was, he had.

CHAPTER TWO

To Cowley, and to War

I have never been guided by anyone. I have always played off my own bat. I have listened to advice from bankers, auditors and solicitors, and still gone my own way when I thought they weren't seeing far enough. I have done many things altogether against engineering principles. That's how the car was made, and that's how it grew…

In an interview, 1926

The year in which Morris had decided to work for himself, making bicycles, was also important for another man, an American fourteen years his senior: Henry Ford.

As a power-house engineer in Detroit, Ford had spent every free evening and weekend for seven years constructing a vehicle that would run under its own power. Like Morris, he started from scratch. He took wheels from two old cycles, the wheel hubs from old railway washers, and made the cylinder from a piece of steam pipe.

His car was a tiny, frail thing, like a pram, even after all his work, but it had the ultimate virtue: it worked, and one night in 1893, this first Ford car carried its designer round the streets in a rain storm. His wife stood outside their house with her umbrella up, listening for the beating of its engine, hoping her husband was safe.

While Morris was still repairing the cycles of others, Ford had talked the Mayor of Detroit into giving him permission to appear on the streets of the town in his car in the day time. Not all his appearances were successful. Sometimes, when the little engine failed him, he would padlock the whole car to a lamppost lest anyone should steal it, while he ran home for spares. But by the time Morris decided to build his own car, Ford was already a millionaire, risen to riches on the triumphs of his Model T, which even in those days was being exported to Britain in considerable numbers. They were tall, unsightly cars, the butt of music-hall jokes, but they ran. A Mr Henry Alexander drove one four thousand feet up Ben Nevis in 1911, and after this Ford had no need for further publicity stunts, for in that year he sold 14,050 cars.

The two-seater Ford was selling for £135 just before the First World War, and the four-seater for £150; and to have the car converted from left-hand to right-hand steering cost about £12 more. This was substantially under the price of any other recognized make of car, but nevertheless many people would have been willing to pay perhaps £50 more for a British-made car and one that was not the subject of so much good-natured fun. This Morris knew, and he determined to produce such a model.

At the motor show in 1912, he met Mr Gordon Stewart, who had a motor shop in Woodstock Street off Bond Street, with a window big enough to take two cars.

Morris showed him his plans for a car.

"Come back with me to Oxford and have a look at it yourself," he suggested. "You can see what it looks like, for I've got one already made. All I want now is to get some orders for others so that we can start production properly."

Stewart came back, saw the car, and liked it. He and Morris took it out for a run and reached Stokenchurch, about twenty miles away on the London road, when the cast iron universal joint broke under them, and they coasted to the side of the road.

No matter. Gordon Stewart liked the car and agreed to take the first few hundred made and distribute them. In fact, he took thousands, and then hundreds of thousands, for the firm of Stewart and Ardern became and remained Morris distributors, and when Gordon Stewart died in 1952 he was a rich man, and a man of many interests. He had helped to found the Royal Society for the Prevention of Accidents; he had pioneered new methods of chicken farming on his own farm at Send Manor, Ripley; and he owned the Strand Theatre, which he bought in 1946 for £150,000.

His prosperity was very closely tied to the prosperity of William Morris. In the first year after the 1918 Armistice, for instance, Stewart's turnover, with a staff of six, was £32,000. By 1923 – then employing three hundred – it had leapt to more than a million pounds; a success that sprang from a moment when Morris had pulled the sketch of a car from his pocket and Gordon Stewart had realized that this was not just an ordinary blue-print, but also a plan for the future…

Incidentally, although it was humiliating for W R M to have his car break down like that on its first important journey, he learned something from the incident. On all his cars thereafter he used steel for the universal joints instead of iron, and he enclosed the drive from the gearbox to the back axle so that dust and grit from the road could not harm the moving parts.

Well, there was the car. Now to announce it.

In the *Autocar* of 26 October 1912, there appeared this intelligence:

THE MORRIS-OXFORD LIGHT CAR

A new miniature light car has been put on the market by W R M Motors Ltd., of Longwall Street, Oxford. It is nominally of 10 hp and has a 4 cylinder White and Poppe water-cooled engine of 60 mm bore x 90 mm stroke with enclosed valves on opposite sides of the engine, a White and Poppe carburettor, and a Bosch magneto.

The crankshaft has three bearings, the valves and ports are large, and adjustable tappets with fibre insets are provided…

Lubrication was by automatic pump, through the hollow crankshaft, and there was an indicator, in the form of a glass tube on the dashboard, so that the driver could actually see that the oil was in circulation and know that it was reaching the bearings.

The multiple-disc clutch ran in oil, and the engine drove through a three-speed gearbox, with a torque tube; final drive was by worm; steering was by worm and wheel.

The brakes worked on the back wheels only, in the fashion of the time, and the wheels were detachable, 700 x 80 mm. The springs were unusually long for the size of the car, which was 10 ft 5 in from bull-nose bonnet to its tail. The wheelbase of this first Morris was 7 ft; and the track, 3 ft 4 in.

Equipment was comprehensive: Cape hood, adjustable windscreen, two Powell and Hanmer acetylene headlights with Mangin mirrors, and paraffin side and tail lights. Each car left Oxford supplied with a horn, pump, jack, tools, and a spare wheel that had no tyre.

The *Motor* was equally enthusiastic. On 17 December 1912, Walter Groves wrote: "A very shapely external appearance has been obtained, the radiator being of the prow shape... The body is a two-seated flush-side torpedo, painted pearl grey, upholstered in leather and brass mounted...in every sense a high-grade production..."

At £175, this newcomer was cheap enough and attractive enough to have a very wide appeal. It cost £40 more than the two-seater Ford, and £25 less than the Turner Ten. Right from the start, the Morris was an in-between car; a car for the man who did not have to buy the cheapest, but who could not afford the most expensive: the perfect car for the middle-class.

Because he had made the car himself, and in Oxford, W R M called it a Morris-Oxford; and when he moved out to Cowley and started to make a cheaper version of the car, he called that the Morris-Cowley.

As a badge for the front of the radiator of every car he made, he chose the crest of Oxford City. This shows an ox, tail up, crossing wavy lines that represent a ford over the River Thames. This crest has become familiar on hundreds of thousands of Morris radiators, and on Morris dealer signs round the world, and indeed has become symbolic of the new industrial Oxford which grew up largely as the result of the growth of the Morris enterprises. The day William Morris set up his business to make motor cars was the beginning of the end of Oxford as the quiet city it had been, and the time was coming when the University would be referred to a little sadly by its own members as "Cowley's Latin Quarter".

Nowadays, it is the custom for those who mourn this passing of an Oxford they knew to say that Morris should "never have been allowed" to start, but at the time no one ever imagined how the business would expand. Many owners of small garages up and down England began to make cars at this time, too, without any such success; there was only one Morris.

In 1912, casting about for some building where he could assemble his cars, he was offered the use of an old military college, off the Cowley Road, part of which was an extension of the grammar school where his father had been educated. A smallish manor house adjoined the old buildings, and he and his wife decided to live there, near his work. Today, their original dining-room table and chairs are still in service in the old boardroom at Cowley.

The college possessed a hall, a little longer than a cricket pitch and about thirty feet wide, and here he began to make his cars.

"It was an adventure," he admitted later on, "but I felt that if I worked as hard as ever, success was bound to come…"

The factory was a cold, draughty place. They did their best to keep it warm with braziers and oil-stoves in the winter, but there was not much time to feel the cold, for they were too busy working.

In America, when Henry Ford started to build cars at Highland Park the system was adopted of dragging a chassis across the floor of the factory at the end of a rope. The mechanics stood on either side of the rope, near small piles of components, and as the chassis moved along between them they fitted the wheels, the body, the engine and transmission. Thus was born the assembly-line method of car production.

Morris did not use a rope, but he employed the same principle, where each man had a set job to do. Ford used to say sourly, "From time waste there can be no salvage. It is the easiest of all waste, and the hardest to correct, because it does not litter the floor." There was little time wasted at Cowley.

"Commonly, I toiled for thirty-six hours at a stretch," Lord Nuffield recalls now.

He made no components at Cowley, only cars. He bought all the parts elsewhere and assembled them in the big hall, but right from the start, Morris insisted on a high specification for every part he ordered; there was no shoddy stuff allowed, and no shoddy work, either.

He wanted his cars to be cheap to run, without the need for frequent repairs or attention, and this meant that all the parts had to be of good quality and well designed. The fact that some of these first two-seater Morris-Oxfords are still running proves they were.

Morris bought the engines and gearboxes from White and Poppe, of Coventry, where Burgess had been employed; the axles and steering from Wrigleys, of Birmingham; the frames from Redpath, Brown; the lamps from Powell and Hanmer; the wheels from Sankey, and the tyres from Dunlop.

Hollick and Pratt, a long-established Coventry firm of coachbuilders, made the bodies, and these were delivered from Oxford station by horse and cart. Later, years later, the car that Morris made in Cowley had a unique distinction: a railway goods station called, after it, Morris-Cowley...

One of Lord Nuffield's employees in these days was a young lad, still in his teens, named George Lucas. Now, like W R M he has prospered and been raised to the peerage as Lord Lucas of Chilworth, but still his memories of the early times remain fresh and warm.

"They were great days," he recalls nostalgically. "We were all part of a team... Morris knew how to get the best out of each one..."

Morris' brother-in-law, Bill Anstey, who became Transport Manager, a position he held up to the time of his death in 1941, worked in a loft above the hall, painting the wheels for the cars. As he finished each one, he would lower it down slowly on the end of a rope.

The cars were supplied in two stock colours: grey, a colour that had always appealed to Morris, and black. They could also be supplied in other colours to special order. The wheels were painted either grey or blue. They would call up to Bill Anstey, "Another set of blue wheels", or, "A set of grey", and down they would come on the end of the rope.

Nowadays, when the men working in the Tyre Bay at Cowley can fit as many as 320 tyres in a single day, and when 1,100 gallons of paint are used in the same time, all this seems very primitive. But although mechanization to an almost ultimate degree has improved and accelerated the way in which Morris cars are built, so that from the four main production lines at Cowley, which are all a quarter of a mile long, twenty vehicles can be produced in one hour, still the basic idea behind it is the same.

Into the main stream of the assembly plant at Cowley there flow other lesser tributaries: the supply of engines and bodies and wheels. These are carried by mechanical conveyors, moving belts or overhead rail arrangements, and never can the supply be broken, for then a hold-up in production would result and that, in a factory devoted to efficiency, is the unforgivable sin. The cars now made at Cowley are built on the "mono-construction" principle, which means that the body and chassis are one unit and not two bolted together. The body shells at Cowley first move on a conveyor belt to a gigantic tank called the Rotodip where, like chickens on spits, they are rotated in clamps and dunked in special rust-proofing fluid, and then coated in primer paint. This huge plant, 306 feet long, can clean, make rust-proof and coat with primer paint, twenty car bodies every hour in one continuous series of operations. Then, by overhead crane, they move to the Paint Shop for spraying in their undercoat and final colours; then the axles are fitted, and the body shells are lowered on to a conveyor for the engine to be added, an operation that only takes eighty seconds.

Wires are connected to the lights, to the instruments and the other electrical accessories; wheels are automatically deposited by the side of the cars; then follow the seats and upholstery, and inspections, checks, and all the many operations that go to form the great amalgam of effort and genius that is the modern motor car. A long way from the days of the drill hall in 1913, agreed – but everything has to begin somewhere; the acorn must always precede the oak.

"There were no power tools in my day," Lord Lucas recalls. "The nearest we had then was a hand-drill, but still the cars were made quickly, and they were made well."

Mr Gordon Stewart handled the sales in London, but sometimes local orders would come in, and then Morris would put on a dark suit, gloves and a bowler hat, and act as his own salesman. Frequently he delivered his own cars all over the country, returning by train through the night. He seemed tireless, a Trojan in shirt sleeves, smoking almost incessantly, using a holder so that the smoke kept out of his eyes; the first to arrive at the works in the morning, and the last to leave at night.

The office-work was done by one girl, and the accounts and books were kept by Frederick Morris, at Longwall, who shared the enthusiasm of his son and had joined him in his venture. They worked together until Mr Morris senior died in 1917. Although he never knew what it was to draw up a cheque for a million pounds for his son to sign and give away to charity, he had seen him spared for the first lap of his tremendous journey; he had seen the end of the beginning.

Lord Nuffield has always been generous in giving credit for his success to both his parents. They believed in him, and their belief was justified. His mother was eighty-four when she died in 1934, the year her son was created a peer for "public and philanthropic services", and all her married life she stayed on in her little villa. Although time and again Morris had offered her a fine new house anywhere she wanted, and a staff to fill it, she was quite content where she was, and could not be persuaded to move.

A year after her death, Lord Nuffield unveiled a bronze likeness of her in St Peter's Hall – to which foundation at that date he had given £10,000 – and where one of the buildings is named after her. Every day of his working life, in peace and war, prosperity and worry, he had gone to see his mother, and in death she was still remembered. As he unveiled the memorial plaque to her, he said, "I am glad to know that her face, as I remember it as a young man, will become familiar to generation after generation of young men as they pass through this door...

"As they look, they will understand why her son wishes to perpetuate her memory, by seeking to help his younger brethren whose ambitions were similar to his own, when his mother's love and care were the mainstay of his life..."

Like so many other famous men he knew that his debt to his mother was incalculable. He paid it in the only coin a mother ever asks for: the currency of love returned and triumph shared.

Lord Nuffield has been equally generous in praise of his father. "My father indeed had riches, but of the mind, not of the pocket," he used to say. "The least valuable thing a parent can endow a strong, healthy son with, is money. Counsel, correction and example should count far more in equipping him for the battle of life.

"It is no regret to me that I was not the son of a rich man..."

In his first year as a full-time motor manufacturer, Morris remembered how valuable publicity had been to him when he was making cycles, and he was sharp enough to turn an incident in local politics to the advantage of his firm. Oxford City

Council had decided that their city should have some form of public transport: either motorbuses, which were in use in London, or tram-cars running on rails, which were also common in many other cities.

Some people wanted one and some the other, but it was generally believed that while the Council of the day wanted trams in the streets, the citizens wanted buses. Certainly, the college authorities could not have been very keen on the idea of trams grinding and swaying down the Cornmarket or the Broad as they did along the Old Kent Road, but it seemed as though this would actually happen. And then W R M took a hand.

Out of the profits of his business he acquired some Daimler buses, hired drivers from the old London General Omnibus Company to drive them, and early one morning, and quite unexpectedly, they appeared on the streets of Oxford.

People were astounded; most of them had literally never seen anything like this before, and crowds turned out to watch the buses rumble by on their solid rubber tyres. The police also turned out in numbers, and they stopped each bus and asked the driver and conductor under whose authority they were running.

Morris had anticipated this move, for it was never his intention to take money from passengers, as that would be infringing the law and he could have been liable for prosecution. Instead, he had arranged for free tickets to be available at certain shops in Oxford. Thus there could be no prosecution, because technically there was no offence!

For the people of Oxford the whole affair became a most entertaining experience, but for their Council it was an insult, and a very grievous one. As they could find no legal way of stopping Morris, for, after all, there is no law that prevents a man running a bus at his own expense and giving people free lifts in it, they decided to take him on at his own game and run buses themselves.

Morris went one better: he ran *two* buses, one before and one behind the Council vehicle.

The whole thing became a farce; three buses trundling through the streets, all of them nearly empty, and always the Council bus the emptiest of the three, for who would pay when a free seat was available?

The Council threw in their hand.

They decided they would run buses instead of trams, and as Morris already had a fleet, they made an offer for them all. He accepted it, with a thankfulness that only he knew, and which he did not reveal for many years. Then, looking back on the episode, he recalled with a grin, "It's just as well they bought them when they did, for if we had gone on much longer, we'd have been forced off the road ourselves – broke!"

This was an admirable diversion for Oxford, and one that kept trams off its lovely streets, and it was also important for Mr Morris at Cowley. He became "talked about". He was news.

"What's this man Morris up to?" people asked, and with cause, for this man Morris was up to great things. Between 1912 and the outbreak of the First World War, he made and sold four hundred Morris-Oxford cars – a tremendous number for those days.

"I remember yet the excitement when we hit our peak production – thirty cars in one week," says Lord Lucas. "That took some doing, for Morris didn't only make 'em – he sold 'em…"

The speed at which Morris was making and selling his cars was nearly his undoing, for he began to have difficulty in buying the components he wanted in the numbers he needed; and he found it hard to impress on the makers of these parts the importance of making them all to exactly the same pattern.

In those days, car manufacturers worked on each vehicle as if it were the only one, with the result that no two cars were ever precisely the same, even though they might be the same type. Often the manufacturers made the parts themselves, then machined them and fitted them together with a care and a craftsmanship only equalled by its expense. They made so few cars that no one thought of copying Ford across the Atlantic:

working out a sequence of jobs to be done and doing them in a methodical order, so that instead of making cars in twos and threes they could make them by hundreds and thousands.

The cars of the British manufacturers would be half-built in one part of their factory, and then they might be wheeled over somewhere else for other operations, such as wiring or tuning, and then pushed into another shed for the upholstery to be fitted, and so on. Such ways of building seem amateurish today, and are known as "knife and fork methods". Each car made like this is an individual vehicle in that there is no other quite like it in every specific detail, and while this method of construction may succeed where the ultimate cost is of no special consideration, for Morris, who wanted to produce a good article at a low price, it was out of the question.

He wanted to buy parts by the gross, all made to the same standard and of the same quality, so that a car could be assembled from them as a boy makes a model from his Meccano set. Then, if a part wore out, it could be easily and cheaply replaced; but he could find no firms in this country in those days that could guarantee to supply him in the quantities he wanted.

About this time, advertisements for a car called the Pick were appearing in the motoring magazines. In one of them the makers, the New Pick Motor Co. of Stamford, Lincolnshire, stated boldly:

> We sell *direct* to the Public at Rock Bottom nett Cash Price. If an Agent tries to crab the Car it is because you can buy it at the same price as he can. On other cars of this class (the two-seater sold at £150; the four-seater at £165) the Agent's commission would be at least £40 which, of course, the buyer indirectly pays.
>
> Don't be led away to buy a Perambulator or a Runabout when you can get an excellent *British made car* at the above price...

An individualist called John Pick was behind this venture, and the cars he made were good. At one time, indeed, they were a serious rival to Morris cars, but then the war came, and Pick had to spend a great deal of money keeping patents alive, and although he restarted the manufacture of his cars after the Armistice, he could not meet the hardening competition of the post-war years, and soon he gave up his business. He decided to enter market gardening instead, which he did with some success, working until shortly before his death at the age of eighty-five in January 1954. He grew and sold his produce from a shop on the Great North Road, and every day, as Morris cars rolled by, he must have wondered whether his name would have become as famous as that of Mr Morris had he persevered in his endeavours with *his* cars.

The owners of the foundries in Coventry and Birmingham that supplied other motor manufacturers with their components were also not keen to change their ways of business at the whim of a young man writing from a village in Oxfordshire. If he wanted that type of work, they said, he would have to go elsewhere for it. In the end, he did.

He went to America, and he took with him another man about his own age called Hans Landstad, who had helped in the design of the first engine for the Morris-Oxford. Young Landstad, who retired in 1947, was a great character. He was the son of a Norwegian pastor, and as a boy he ran away to sea and sailed before the mast to Australia. Then he came to England, and used the experience he had gained with marine engines to help him in a career with a Midlands engineering firm.

Morris met him in Coventry, and it was a meeting of kindred souls, for although neither could understand the language of the other perfectly, and sometimes they had to fall back on dumb show and manage with the aid of sketches hastily drawn on pads, they could recognize each other's worth, and Hans Landstad was a genius with an engine.

Morris and he sailed to America, where other manufacturers had already learned much from the mass production efforts of

Henry Ford, and the parts the two men ordered were delivered in Cowley on time and in the quantities they asked for.

In 1914 the Morris-Oxford was improved in several particulars. The wheelbase was increased to 7 ft 6 in to allow room for a dickey seat, and the track widened to 3 ft 8 in. The rear springs were underslung, as they remained for many years, and the shackle pins of both front and back springs were of a larger size than had previously been fitted. The steering wheel rim of the 1913 car had been of wood; for 1914 it was of metal, covered with black xylonite. Also, the connecting rod from the steering box to the axle steering arm was put above the box instead of below it, where it had been vulnerable if the car was taken on rough tracks.

In a standard green colour, with a fine white marking line, the new car was priced at 190 guineas. The 1913 model cost £180.

The *Autocar* was full of praise for the design:

"This newcomer to the ranks of miniature cars... undoubtedly created a most favourable impression on account of its excellence of design", wrote one correspondent.

"Without attempting to enter the field of the cycle-car proper, it aims at being essentially a miniature motor car, and such indeed it is, possessing all the attributes of a full-size car, but – and here is where it differs from many of the small cars on the road – it is also made with all the care that is bestowed upon the highest priced car...

"There is more of the Rolls-Royce *cachet* about it than most wee cars possess; and in general reliability, the silky ease of the steering, the velvety efficiency of its clutch and the operation of the brakes, is simply astoundingly good."

A coupé on the Morris chassis was introduced by Hollick and Pratt in the same year, to sell at £245, and Morris was anxious to increase his production so that he could lower the price of his two-seater to £165. The annual road tax at this time was £3 3s.

One of his latest rivals was a two-seater called the Taunton, that sold at £171 without either windscreen or lamps, and looked very like the Morris-Oxford with the same bull-nose radiator. Behind the driving and passenger seats, however, was a third crude seat, in wood. This car made its debut at a very unfortunate time, in May of 1914. The makers took pride in announcing that "among the well-known people who have already ordered Taunton cars" was Field Marshal Earl Kitchener. By August, however, the Field Marshal had other and more important things to occupy his mind. War had been declared, and the Taunton went out of production.

So did the Morris-Oxford, and for the duration of the war the little Morris factory at Cowley saw more brass hats and military red tabs than had ever been there before, even in the days of its glory as a military academy.

CHAPTER THREE

The Car you Buy to Keep

When things are difficult, do not moan, but set to and get business. It will not always come to you.

One of Lord Nuffield's maxims

Morris has always been tremendously patriotic; sometimes almost boyishly so. In the twenties, for instance, he had an enormous Union Jack flown from a mast above his factory at Cowley, and floodlit at night so that it could be seen for miles around. Later, he used the slogan, "Buy British and Be Proud of it", and then dropped this in favour of another appeal to the patriotism of the buyer: "Even if you don't buy a Morris, at least buy a car made in the United Kingdom." Always he preached, and believed, the gospel that what was British was also best.

One day in the mid-thirties he saw three undergraduates outside the factory gates distributing Communist leaflets.

"If Russia is such a grand place, then why don't you go there?" he asked bluntly. "I'll pay your fares myself. If it's so much better, go and see it, but there's one condition. You'll have to *stay*."

The students were amazed and startled by this offer, but not so much that they accepted it. Morris snorted his contempt for people who professed to believe in something and yet had not the courage to put their theories to the test... That is the sort of person he is: plain, blunt, forthright, and immensely loyal to his own country and his own friends. Thus it was entirely in character for him to try to join the Army in August 1914, but the CO of the Oxfordshire Light Infantry turned down his offer of service.

"Go back to your factory," he said, not unkindly, when W R M reported to him. "You can do much more useful work there."

So, indeed, it proved.

The little factory with its score or so of men was taken over by the Government; Morris was appointed the "Salaried Comptroller of the Trench Warfare Dept, Cowley, Oxford," and they began to produce war material.

Looked at in the light of the tremendous efforts of the Nuffield Organization in the Second World War, when sixty-three factories were under their direct or indirect control, and more than thirty thousand workers building tanks, lorries, guns and torpedoes, and repairing nearly eighty thousand fighting aeroplanes, the output of the Morris concern between 1914 and 1918 was very small; but their efforts were large in proportion to the size of the place.

At first they produced shells for Stokes Trench Howitzers, and then mine-sinkers for the North Sea minefields. These were large, square metal boxes, packed with clockwork machinery, and used for regulating the depth at which mines would float in the sea. In four years of war, Morris and his men made fifty thousand of them, and in 1917, in recognition of his services, he was awarded the OBE.

The works spread out round Cowley, and a railway siding was run in from the main line. This was going to prove useful afterwards.

What spare time Morris had he spent thinking about the car he would build after the war. He circulated friends and the editors of motoring magazines as early on as 1915 with rough plans of his intention for their comments. He realized that there would be a boom in spending and that many more would be able to afford a higher standard of life than they had dreamed possible before the war. There would be a demand for goods that had previously been reserved for only a very few.

When the bells rang out for peace, however, and when the crowds filled the streets, cheering the news of the Armistice, he looked about his factory and wondered whether all his efforts had, after all, been worth while. He was tired, and so were his men. They had worked at their limit for years. Their machinery was old and run down, and there was little money to his credit at the bank. Unlike many other manufacturers during the war, he had worked for a set salary and not for the far greater profits he could have made.

Could he make a living out of making cars again? There was only one way to find out, and that was to build them and see whether they sold or not. And this he proceeded to do.

He was forty-one, an age when many employers consider a man "too old" to be taken on the staff; an age when physicians claim that the machinery of the human body begins to slow down. But Morris was not too old at forty. Like Shaw and Fleming and Churchill and a dozen others, he was still too young; his real life's work still lay ahead of him.

Ford was off to a long start, with a factory in Manchester where the cars were assembled, and thousands of post-war Model T Fords were already running on the roads.

Morris knew that Henry Ford was already a millionaire, while he himself was not even well-to-do, but he had the same tiny seed of genius within, and it did not fail him. He knew he could do what Ford had done; could equal it; and, having adapted it to British conditions, could better it. This he set out to do – and he succeeded.

It was hard work. Prices of all the components he had used in 1914 were up two or three times on their pre-war figures, and so were wages. But he found one virtue: the war had taught Midland manufacturers something of the value of mass production.

The French company of Hotchkiss et Cie, for instance, who were famous for their machine guns, had established a wartime branch factory in Gosford Street, Coventry. They were at this time more widely known for the armaments they made than for the Hotchkiss cars, which had the firm's badge of two crossed cannon on their radiators. They agreed to make engines for Morris in quantity. Other firms, too, were more willing to supply components in bulk than they had been five years before; but for all that, starting was a slow job.

There were delays. The returning servicemen were out of touch with conditions, and there was some bad feeling between the ex-soldiers and those who had stayed at home working in war factories. No one really knew what the future held. Some wanted to return to the old Edwardian days of grace and gentle living; others clamoured for the new world for which they felt they had been fighting. There was still a lot of money about, however, which was just as well, for Morris found that his pre-war selling price would be nearly doubled. The post-war Morris-Cowley, with a wheelbase increased to 8 ft 6 in, and the track widened to 4 ft, with electric lighting but no electric starting, now cost £315. The Morris-Oxford, with the same engine and measurements, but more comprehensive equipment which included an electric starter, cost £360 for the two-seater, £390 for the four-seater, and £450 for the coupé.

Even at £315, however, the Cowley cost substantially less than some of its competitors. The 11.9 hp Meteorite cost £380 for the bare chassis, and the 10-12 hp Mercury, £375. On the other hand, a light car called the Palladium, a two-seater with two cylinders, only cost 265 guineas, and the flat-twin 10 hp ABC was advertised for as little as £195. These two cars were not nearly so solidly made as the Morris, and they had not the pre-war reputation for reliability. They were brave attempts to capture the market, but in business bravery is not enough. Both of them soon died.

There were many other brave and ingenious attempts to make a market for a new car or a novel accessory. The 13–30 hp Spyker had a small mechanical air-pump to blow up its tyres, fitted in front of the engine and operated by a hand control. The HFG was started by a pedal on the floor near the steering column. The Binks sparking plug had an insulated top which could be lifted and the terminal held close to the body, "thus showing the quality of the spark". Petrol in 1919 was expensive, and there were many devices on sale that claimed to give economies. For the most part these consisted of an extra valve in the inlet manifold which could be opened to weaken the mixture. Similar devices are marketed today.

Among cars, the Rolls-Royce occupied the same position then as now, but two American cars copied its classic radiator without being able to acquire its breeding. They were the 29.4 hp Roamer, and the Moon. "Don't sigh for the moon – you can get it for £775 – the Victory 6-cylinder Moon," said the makers.

All motor manufacturers used their wartime experience to the best possible advantage. The war was still something to remember with pride; the disillusion came later, with *Journey's End*, the brass bands of unemployed ex-soldiers in the gutters, the miners marching and singing through the rain from Wales to London. In 1919, anything that recalled the war was still treated with respect; it had been an affair of honour.

Thus, the HFG was "made like an aeroplane", and the Crossley was described as "the car that's served the RFC in every battle area". Crossleys still called their 25–30 hp tourer "the RFC Model". The price, unfortunately, militated against the widest popularity. The chassis alone cost £850.

Ex-pilots fitted old aircraft instruments to their cars, and much in demand was the Gabriel Bugle – a horn which, so it was said, "makes a musician of your chauffeur". It was really a tiny four-pipe organ worked by the exhaust and "played" with four keys.

Many engineers were producing small cars, for the market seemed immense; and so it was, but not large enough to support all the new makes of car that vied with each other for the public favour. The Angus Sanderson, the Arrol-Johnston, the Belsize, the Cluley, the Cosmos…the Cubitt, the Chiltern…the Thor, the Eric-Campbell, HE, GWK, Varley Woods, Waverley…all gone now these many years; and most of them forgotten. Many of these cars were well designed and made, but not all the firms producing them could weather the economic storm that blew up in the early twenties. They had no reserves of cash to fall back on; they went under and were heard of no more, although some of the cars they made are still in use today.

Morris cars were being advertised in an unusually idyllic way at this time.

"A clear spring morning, with long shadows on the grass…a hamper of sandwiches, a camera and a friend…a day in the country. Tea at the old inn; the gathering dusk; the night with a million stars and a restful content as you glide homewards" – and an arrow pointed to the driving seat of a Cowley two-seater.

Then the dusk gathered indeed, and in 1921 the car makers discovered with alarm that their cars were not selling. They began to cut down production and to stand off men. Morris had his drill-hall at Cowley full of cars he could not sell, and all over the country it was the same: cars in the showrooms and only the salesmen there to look at them. Clearly, something had to be done, and done quickly, before it was too late. But what to do? If you couldn't sell the cars, you couldn't. Costs were still going up. Food was dear; so were the engines, the tyres, the springs. Reluctantly, many manufacturers put up their prices. It was either that, they reasoned, or they would be selling their cars at a loss. What they did not realize – or dared not realize – was that for the most part they were not selling their cars at all.

Many firms that only produced a small number of hand-built cars every year, had calculated their profits on the basis of a fair sum being made on each one, and they could not survive the slump. Most ran for what financial shelter they could find, and Morris was urged to do likewise.

"Cut your production, cut your staff and your overheads, man, or you'll be ruined," his friends said. "You'll lose everything you've built up. You'll be worse off than you were when you started building bikes, for you'll have nothing…"

There was wisdom in this, but it did not appeal to W R M. He believed in the maxim that attack is the best method of defence.

In 1921 there seemed to be two courses open to him. He could do nothing, and hope that somehow, some time, prospects would brighten and his cars would sell. Or, if no one would buy the cars at their list prices, he could offer them for sale at much less, and hope for a tiny profit on each one.

As he saw things, he had nothing much to lose. If the gamble failed, then he would go broke no more quickly than if he had done nothing at all. Since, in his view, any sort of action was better than just waiting for the eventual catastrophe, he decided to cut his prices.

In the *Autocar* of 12 February 1921, there appeared this announcement:

> The prices of Morris-Cowley and Morris-Oxford cars are reduced as below:
>
> Morris-Cowley 4-seater down £100 to £425
> Morris-Cowley 2-seater down £90 to £375
> Morris-Cowley chassis down £65 to £325
> Morris-Oxford 4-seater down £25 to £565
> Morris-Oxford 2-seater down £25 to £510
> Morris-Oxford coupé down £80 to £595

"These reductions", Morris added, "are coincident with an improvement in the already excellent Morris-Oxford model, and represent an attempt to cheapen the price of a very sound British type of car by increasing the demand..."

The attempt was fantastically successful, as a few months showed, but at the time not every other car manufacturers agreed that this was the best way to attract new business. Several thought it was the worst possible way to go about things. The Star Engineering Company, for instance, who made Star cars, announced that they would indemnify all purchasers of their cars against a possible reduction of prices up to the 1st of July of that year. The cheapest Star was the 15.9 hp model at £790, and the dearest, the Touring Car, at £1,065.

The makers of the Deemster light car decided on a similar course, and they promised all who bought their 1921 models, at £475, that there would be no price reduction during "the present season".

Some other firms, however, realized that Morris was on the right track, and they hastened to cut their prices quickly, or to offer other inducements to buyers, lest he should capture the entire market. The 10 hp four-seater Citroen came down from £495 to £395, "the equivalent of 600 gallons of petrol FREE – enough on the Citroen for 23,000 miles, or once round the world", as the makers pointed out with pride; and Hillmans were all guaranteed for twelve months and seven days against "mechanical and electrical breakdown", and the cost of car hire up to £30 was promised to their owners during repairs.

Morris was first, however, and he received liberal credit for his initiative. The moment when he decided on the bold stroke of getting rid of his cars almost at cost price was the real foundation of his fortune.

In October, 1922, he cut his prices again, so that the two-seater Cowley was on sale at only £225.

"Surely you're not serious?" colleagues asked him, when he suggested this move. "It can't be done." But it was.

"W R had guts," was the way one of his own workmates put it. "He decided that he might be hanged as well for a sheep as for a lamb, so he marked down every model. The results were fantastic..."

They stayed that way all through the twenties and thirties, with W R M making more money more quickly than almost anyone else in the country. He was learning the lesson that Ford had learned before him with cars, and Lipton with tea, and Woolworth with many things: cut your profits and your business will boom.

When success came to W R M it did not bring carelessness behind it. He still paid attention to little details of design that others tended to ignore. For instance, he enclosed the front and rear springs of the 1922 Morris-Oxford in leather gaiters so that dust and grit and mud could not harm the leaves, a provision that few cars have, even today. Then he watched how buyers reacted to his models. In 1923, when the Morris-Oxford was costing £446 10s, one customer wrote a letter to the company: "I cannot but feel that at this price the equipment should include a speedometer and an electric horn. I rather missed both these accessories which are a necessity these days…"

Morris agreed, and next year his cars had both these fitments, and for the year after, he included a comprehensive insurance policy for good measure. He was proving that Ford's maxim was perfectly true: "A business absolutely devoted to service will have only one worry about profits – they will be embarrassingly large…"

In the early nineteen-twenties, on every car imported into Great Britain from abroad, save from the Empire, there was an import duty of one-third of its value. This was called the McKenna Duty, and it was a most important safeguard at the time, for the British motor industry, catering for a relatively small number of motorists, could not hope to compete with American manufacturers who could sell their cars much more cheaply because of their enormous output.

In 1924, the Government of the day decided to repeal the duty, a move that was viewed with alarm by Morris and the other car manufacturers because, in the previous year, Ford of Detroit had produced two million Model T cars. If these could be shipped to England they could easily undersell every British make, and this might mean the ruin of British firms, Morris included.

"Every foreign car on the roads means another British workman unemployed for twelve months," he would say bitterly, and he led a campaign urging that this duty should still be paid on foreign cars. Thanks largely to such strenuous efforts by Morris and his contemporaries against its removal, the McKenna duty was put back very shortly afterwards.

"If it hadn't been for these American cars," Morris would say later on, "we'd have captured the entire British market." Even as things were, they captured very nearly half of it.

In 1919, Morris made 337 cars; in 1922, 5,166. For the 1923 season his turnover was £6,000,000, and for the year after, nearly double. That was the measure of his success, and it was only the beginning.

There were many other makes of car that were as soundly designed as the Morris, but only a few still survive. How has the Morris remained so successful a car, for so many years?

There are several reasons, and probably the main one is that they have always been constructed with a definite market in view. There has never been anything haphazard about that. Morris built them for the middle-class family man; they were cheap to run, and reliable, and they also had character. They have always been, in their class, the right car at the right time.

Other makes that have not survived, like the Moller, the Newey and the Motobloc, had too limited an appeal for a wide sale. The 12 hp Motobloc four-seater, for example, cost £895, and although people might gaze in admiration at them through the showroom windows, they could not afford to go in and buy them.

Between 1922 and 1923 Morris sold 15,987 cars; and then in the following year he cut his prices again and sold nearly double this amount, so that by December 1925 he was selling nearly 50,000 a year. They were advertised as "The Car You Buy to Keep", and considering their price, their equipment was extremely comprehensive.

The Cowleys, for instance, had the following items of equipment listed in 1923:

Lucas 12-volt dynamotor-starter lighting set (5 lamps), All-Weather hood and side curtains opening with doors, Smith speed-indicator and 8-day clock, Boyce Moto-meter (radiator thermometer), petrol and oil gauges, Enots pump chassis lubrication, 5 Dunlop cord tyres, on 5 detachable steel wheels, spring gaiters, tool-kit, with jack and pump, spare petrol-can and carrier on running-board, and half-gallon tin of Shell engine-oil, painted Morris Grey.

The rather more expensive Morris-Oxfords were equipped with the following:

> Lucas 12-volt dynamotor-starter lighting set, with 5 lamps and dashboard light, Smith speed-indicator and 8-day clock, driving mirror, electric and bulb horns, Gabriel shock-absorbers to both axles, 5 Dunlop Cord tyres (28 x 3^1/$_2$) on 5 detachable steel wheels, Enots pump chassis lubrication, Boyce Moto-meter, petrol and oil gauges, All-Weather One-man hood, 3-panel front windscreen, complete kit of tools, including jack and pump, spare petrol-can and carrier on running-board. Half-gallon tin of Shell oil sent out with every car. Coach-work painted Blue, Claret, Bottle Green or Morris Grey, upholstered in finest mottled grey leather.

Some time after W R M made his first big cut in the prices, when the thunder of praise and the prophecies of woe had died away, he explained how he was making such a success out of his business at the new low prices.

"I looked for my profit from the increased demand," he said. "A net profit of £1 on each car, when I am selling 50,000, is quite a substantial sum."

Henry Ford had much the same reason to give for his own achievement, that same year.*

"I was once asked, when contemplating a reduction of 80 dollars on a car, whether, on a production of 500,000 cars, this would not reduce the income of the company by 40,000,000 dollars," he said. "Of course, if one sold only 500,000 cars at the new price, the income *would* be reduced 40,000,000 dollars –

---

* In his autobiography.

which is an interesting mathematical calculation that has nothing to do with business. Old-time business went on the doctrine that prices should be kept up to the highest point at which people would buy. Modern business has to take the opposite view…"

Morris believed this just as implicitly, and he would add: "I am convinced that success can only be attained by gaining personal experience in every department of one's business, and working out the smallest details for oneself. To manufacture and not to know how to sell is only knowing half the business…"

This was in 1923, and Morris knew not half or three-quarters, but *all* the business. He remembered the early days in the shed at the bottom of his father's garden, when he had done every job himself because there was no one else to help him. He remembered how at his shop in "the High", he would change from overalls, having made the cycle, to dark suit and bowler to sell it, and then into singlet, shorts and skull-cap to race another machine, all to carry his name forward, and to keep it ahead of competitors.

Ford, in the early days of his expansion, used to say: "No factory is large enough to make two kinds of product." Morris was a better psychologist. He realized that the human desire for something different – a better car than the people owned next door – could be used to the advantage of his own firm, and very simply, by making two grades of car.

"The equipment of the Morris-Cowley," as his advertisements for September 1926 explained to the potential buyer, "contains everything that is essential to the practical motorist, but in the Oxford models certain additional items are provided which might be classed as luxuries.

"For instance, on the Cowley, the windscreen wiper is hand operated, whereas on Oxford models it is worked automatically by suction from the engine. The upholstery on Cowley models is of leather cloth. On the Oxfords, it is real leather…"

There was, of course, a bigger difference between the two models, that of their size and the power of their engines. The

Cowley engines were of 11.9 horse power, and the Oxfords, 13.9 horse power, but they were both built on identical production lines, side by side, the cheaper and the dearer, just as Morris cars are made today at Cowley alongside the more expensive Wolseleys; and, at Abingdon, the MG two-seaters and saloons are built alongside the more expensive Rileys.

In each factory, the type of car is the same. At Cowley Morris family cars, small, medium and large, and their rather more expensive counterparts on the Wolseley production line. At Abingdon, both MG and Riley appeal to the man who likes a car for what it can do and how it performs.

Morris cars for 1927 were the first of that make to have Barker Patent dipping headlamps. These were extremely simple in design, and very effective. The two headlamps were mounted on a horizontal bar that ran across the front of the radiator, and this was connected by a system of rods to a lever by the side of the driver. By pulling this he could depress the lamps so that they would direct their beams on to the road just in front of the car, and not dazzle drivers coming towards them.

This mechanical system of dipping only lasted a short time, for Lucas developed a magnetic system whereby the reflector could be mounted on a pivot inside the lamp shell and drawn to one side, left and down, when the dipswitch was pressed, and this killed the clumsier mechanical Barker scheme, for there were no moving parts outside the car to become loose or to rust out. But, for its day and the speeds involved, the Barker system was good and in advance of its time, and that Morris should fit it when he did was an instance of his vision.

Incidentally, this same idea has recently reappeared in a new guise. The little Citroen 2CV has an arrangement whereby the headlamps can be raised or lowered slightly so that they will always be set to their best advantage, even when the tiny car is right down on its springs with a full load of passengers...

In 1927, too, Morris introduced a scheme of fixed charges for repairs and replacement parts, so that owners could calculate just how much it would cost them to have their engine decarbonised or even replaced.

Men who lived through the excitements of the twenties with Morris, recall him in his creased suit, rolling a cigarette from a little machine he carried; bending over plans; supervising new operations; always in the thick of things where the fighting was sharpest, and the going hardest.

Someone asked him, as riches began to arrive, whether he could still take down and reassemble any of his own cars. Morris was rightly amazed at such a naive question.

"Of course I can," he replied. "I should be out of business tomorrow if I couldn't – and do it, I hope, just a bit better than any of my men."

Outwardly, he was as calm as ever, his words as few, his opinions as canny as they had always been. But the grudging admission to a visitor to Cowley in 1926 that, "We shall presently be making a car or two", was not wholly accurate.

That year, in fact, they made 67,000.

CHAPTER FOUR

Expansion

I particularly welcome problems. I like, for instance, to take on a bankrupt concern and turn it into a rattling success. My favourite role is to be the only man on deck when all around have fled – and then to get down to it.

In an article, 1927

Ford and Morris never met. The only time Henry Ford visited Cowley, in the spring of 1928, he forgot to tell them he was coming, and W R M was away at one of his other factories. Ford was shown all round the plant, however, and he expressed some surprise at its size and capabilities.

"You're farther ahead here than most Americans think," he said; meaning, probably, farther ahead than Henry Ford had thought. He was impressed with the energy and impetus of the workpeople. He felt at home in Cowley, which was not really surprising, for both Morris and Ford believed in the same production methods: the need to manufacture in large numbers with a small profit on each one, the necessity to put back the profits into the concern so that the processes of production could be improved, and the desirability of controlling the firms that supplied the main components for their cars so that these subsidiaries could be run as they wanted them to be run.

It was suggested to Morris that, as his business grew, he should make all the parts for his cars in his own factories; tyres, batteries, everything. This he refused to do. He did not want to have the monopoly of such things, but he insisted that every part he used was of good quality, and because he bought on an enormous scale, the prices he paid were low.

The radiators branch of his concern, for instance, was not begun – in 1919 – to force down the prices of those other existing firms that already produced radiators for motor cars, but as a patriotic attempt to help war-wounded and other ex-soldiers who could only do light work. Mr H A Ryder, who later became joint managing director of Morris Motors, was working for the Oxford branch of a Coventry radiator firm, and W R M and he started their factory, the Osberton Radiator Company, together, in North Oxford. In 1926, when Morris Motors became a public company, this became Morris Motors (1926) Ltd., Radiators Branch.

The Morris works grew so rapidly, that some departments were almost continuously on the move, and eventually were half a mile or more away from where they had begun.

As his enterprise grew, Morris absorbed some of the factories that made his components; a process that helped both parties. Morris could keep an eye on the subsidiary concerns, and had the money to equip them on the most ambitious scale; and the firms, for their part, were always assured of work.

Throughout the nineteen-twenties, Morris bought up companies "like a boy collecting postage stamps", as one of his executives described it. If they were solvent when he bought them, he would pump money into them from his own astonishingly successful venture at Cowley, so that with new machines, new methods and fresh heart, their production would soar. If the factories were bankrupt, as they sometimes were, then he would "sort the sheep from the goats"; a process which involved sacking men whose ideas did not coincide with his own, who were old-fashioned in outlook and unwilling to change their views. Then he would set to and make the factories pay again.

Men who had talent and drive, and whose minds were akin to his, he backed with all the resources at his command; and so surprising a gift did he have for selecting such people, even though they might be young and inexperienced, that rarely was he let down. This gift was, undoubtedly, a large part of his genius. He would look at some new man through a haze of smoke from his cigarette in its holder, weighing him up, seeing in his mind's eye not a gauche youngster, nervous in a ready-made suit before a rich employer, but the man he *could* be, backed by enough money, with opportunity to express himself. Other employers took on new hands; Morris took on new brains, men who did not fear defeat because they believed in themselves and knew their ideas would eventually succeed; men like himself, seekers after some ultimate perfection – whether in an engine, or the line of a bonnet, or in the flare of a wing, or in an advertising slogan; men who were never really satisfied because, in this life and world, perfection can never be achieved, but must always be pursued.

Sometimes Morris found them in the factories he bought, as a man who buys a job lot from an antique shop may, in turning it over, find a treasure that has been neglected for years. Sometimes they were already in other jobs, and they impressed him, and so he persuaded them to join his team. They were glad to come, for there was a Napoleonic magnetism about W R M that attracted men of like temperament.

There were times when they disagreed and parted company, for any man of high talent is also a man of individuality; but always they remembered their association with kindness.

In 1923, Morris bought his first outside company, the Hotchkiss concern in Coventry, which had been supplying his engines, and he renamed it Morris Motors Ltd., Engines Branch. Then he acquired Hollick and Pratt, Ltd., the Coventry coach-builders who had made the bodies for his cars, and he called this firm Morris Motors Ltd., Bodies Branch.

Often there was no need, or time, for formal contracts to be drawn up in those early days. Figures, quotations and prices would be scribbled on the backs of envelopes or on old cigarette packets. Morris put his faith in people rather than the written word: a handshake, and a deal was on.

Once, a firm making gears submitted a tender that W R M could not accept.

"I'll bankrupt my business if I cut the price," said the director, when he heard his price was considered too high.

"Well, don't take the order, then," said Morris. But the firm *did* take the order, and within a month the director wrote to W R M announcing that he had made it pay – by increasing the efficiency of his factory.

In 1923 Morris also bought the assets of Wrigley & Co. Ltd., of Soho, Birmingham, the company that had supplied the axles and steering gear for his first car; and on 1 January 1924 the new firm, called Morris Commercial Cars, Ltd., began to produce the first of a long line of lorries and vans.

These lorries were soon in demand. Five years later, several were driven across the Kalahari Desert, "the Southern Sahara", in Africa, the first time any motor vehicle had successfully made the journey. In the previous year, an expedition of American cars had attempted it and failed – a fact that made the Morris attempt all the more valuable from a publicity point of view, for by that time W R M was turning his attention towards the export market.

In July of 1929 he announced his first serious attempt to enter this export market with private cars by giving a banquet for nearly a hundred motor dealers at the Holborn Restaurant. At the end of the meal, as they drew on their cigars and sipped their port, the toastmaster called for silence and Morris stood up to address them. They expected a speech about trade and generalities, with perhaps a leavening of jokes. Instead, W R M spoke briefly about the need to break into the overseas market, the imperative need to export Morris cars. As he neared the end of his short speech he leaned forward and said, "We have built the car to do it, and, gentlemen, *here it is.*"

Two curtains parted at the far end of the banqueting hall, and there, in a platform under the bright glare of floodlights, was a new Morris car, the Isis, which had been designed with a view to export, and even as the diners crowded round the model many Isis cars were being delivered in various parts of the world. The car took its name from that part of the River Thames, at Oxford, which is called the Isis, and it was well ahead of its time. The carburettor, ignition coil and distributor were mounted high up under the bonnet so that shallow rivers could be forded without water stopping the engine. Indeed, during its tests in Wales, the car was successfully run through water two feet six inches deep.

The Isis had been brought to London in pieces, and then assembled in secret in the restaurant. After the dinner, it was dismantled and carried out down the back stairs again, piece by piece. It was a fine car, with hydraulic brakes, pneumatic upholstery, luggage grid, sun visor, bumpers, and exceptionally long springs designed to give both driver and passengers as smooth a ride as possible on the roughest road. Many are still in use today.

In 1933, W R M formed an Export Section near the GWR line at Cowley, capable of dealing with 450 vehicles a week. Now, renamed Nuffield Exports, Ltd., it has handled an average of more than two thousand cars and lorries each week since 1945. Nuffield representatives also work in one hundred and fifty different overseas territories, and to Eire, Denmark, New Zealand, South Africa, India, Holland and Australia, Nuffield vehicles are shipped out in pieces and assembled on the spot.

But we are rushing ahead; such schemes were no more than vague ideas in 1923.

With the Hotchkiss Company, W R M had bought a better bargain than he knew, for on the staff was a young man called Leonard Lord; a tall, slim youth with sandy hair. Lord started his working life in the silk trade, with Courtaulds. Then he had learned engineering as an apprentice with Vickers, at 4s 6d a week for a twelve-hour day. Later, he moved to the old Wolseley company, and then to Hotchkiss.

He had the mark of success on him. By 1933 he was managing director of Morris Motors, Ltd., but three years later, shortly after a disagreement that Lord Nuffield had with the Air Ministry over their plans for shadow factories to produce aero engines, he resigned. Nuffield was not always an easy man to work with. He had founded the firm: he knew the right way to go about things; the others only thought they knew. At the time of Lord's resignation, there was talk of disagreement in the board room, of views that both held strongly and over which neither would give way to the other, but neither of them said so openly. They had worked and prospered together, and even in disagreement they were loyal to each other. Others might liken these two forthright men who would not give way to two railway engines rushing towards each other on the same track, or to the old question in physics: what happens when an irresistible force meets an immovable object? The two men concerned preserved their silence.

The most Leonard Lord permitted himself, when he stepped off the *Queen Mary* after a holiday in America following his resignation, were a few general remarks.

"I am pigheaded," he admitted with a smile, "and Lord Nuffield has his opinions. There was no row between us. A few minutes after we had decided to break our business relations we had a gin and French together, and we laughed over the fact that we could both sit drinking, although we had taken a step that grieved us both…"

Within months, though, they were together again, but not in the motor industry. Lord Nuffield gave £2,000,000 in a Trust to help the distressed areas in Wales, one of his most far-sighted acts, and he wanted a man to administer the Fund, and to report how these places could be helped and their prosperity restored. He asked Leonard Lord to do the job. A year later, however, Lord resigned again, and joined Austin, a firm that had always been W R M's chief business rival.

The little overhead camshaft racing Austins had competed time and time again with the MG Midgets, and the Morris Minor had challenged the Austin Seven. Over the years the battle had gone on up through the horsepower size by size;

Morris Ten versus Austin Ten, Twelve against Twelve; *"I'm going to have a Morris"* ... *"You buy a car, but you invest in an Austin."*

Leonard Lord joined Austin's as their works director, and he looked forward to the rivalry with his old chief.

First, Lord restyled the Austin cars, which had become staid and genteel with the years. He gave them bonnets that opened from the front in the American style, and new radiator grilles.

The war stopped their fierce competition, and over some things they collaborated, as when the Bodies Branch of Morris Motors fitted bodies on to military chassis made by Austins, to the design and specification of the Nuffield Organization; but competition between them began again, undiminished, with peace.

During the war, Leonard Lord had kept a skeleton staff of designers working on ideas for post-war models, and this policy soon bore fruit. For all that, first blood went to Lord Nuffield, whose new Rileys, announced in September 1945, were England's first cars of post-war design. The old Riley dies and presses had been destroyed in Coventry by German bombing, so the company were forced to begin again from scratch.

Then Lord produced his first post-war bid: the Austin A 40, a car that took its title from the brake horsepower developed by the overhead-valve engine. Lord Nuffield replied with a newly-styled Morris-Oxford, using the name of the first car he had ever made; then an MG TD Midget, with independent front suspension and bumpers, designed primarily for the American market; and the new Morris Minor – another return of an old familiar name for a new and revolutionary model.

Austins announced the A 40 sports, an occasional four-seater open car with two SU carburettors and sporting lines in the Italian style; and the Austin A 90, a sports drophead with a hood that could be raised and lowered hydraulically.

From the Riley stable there came a new two-and-a-half litre Riley Roadster, with a wide bench-type seat that could take three people. It had a 100 bhp engine, and a boot that could accommodate a cabin trunk, and was the first open sports Riley to be made since the nineteen-thirties. Austins replied with an old favourite in new dress: the A 30 Austin Seven.

It was a ding-dong battle; point, counterpoint, all the way, both of them enjoying it and both in their element.

Shortly after the end of the war, Leonard Lord went to America to study the market there, and he set up a sales and service organization to deal with the Austin cars he intended to export. It was a wise move, for before the war, the main American criticism of imported cars, and especially those from Britain, was that they had no proper service facilities, and only a few garages stocked large quantities of spare parts. The owners might have their cars off the road for weeks while spares were shipped out from the factories in England.

Nuffield knew what his rival was up to; so he, equally wisely, put £1,000,000 on one side to subsidize any cuts he might have to make in the export prices of his own cars, should the competition become fierce.

In 1948, Lord Nuffield and Mr Lord surprised many of their subordinates by trying to pool their resources; they wanted to stop fighting each other, and fight others instead. The arrangement was not successful. The financial framework around which it was built was complicated; nor, possibly, were the two men who controlled it willing to share and divide their authority.

Then, in 1951, as we shall see later, they tried again, with complete success.

Leonard Lord, knighted in the Queen's Birthday Honours of 1954, was one man who came up with Lord Nuffield, and like him he is an individualist.

There were several others: Miles Thomas, now Sir Miles, and chairman of the British Overseas Airways Corporation, for one.

In the First World War, Miles Thomas was first a driver of an armoured car, and then he qualified as a pilot and won the DFC. After the war he went into journalism and joined the staff of the *Motor*, and then became editor of the *Light Car*. In the early twenties he heard that Morris was going to produce a new model, so he travelled to Cowley to have a word with him about it. W R M went to great trouble to explain all the intricacies of the design to Miles Thomas, but he noticed that his visitor was not making any notes during their discussion.

"Aren't you going to take any notes?" he asked anxiously, for he was used to journalists who scribbled on pads or on the backs of envelopes.

"No," replied Thomas. "I'll remember the facts all right."

"Hmm." Morris did not think much of this; he made up his mind to watch out for any mistakes in the article when it appeared. But when the article was published it was accurate, even down to the most technical details in the design and specification. Morris had been looking for someone to run a new magazine he wanted to start called the *Morris Owner*, and Miles Thomas had impressed him: he offered him the job.

Thomas joined W R M in 1924 with the title of Adviser on Sales Promotion. They were selling three hundred cars a week then; before he left, in 1947, he had seen this soar to a peak of ten times as many. He also made a great success of the *Morris Owner* and, later, of the Morris Oxford Press. Then, when Lord Nuffield wanted a first-class executive to run the Wolseley factory, he made Miles Thomas general manager.

Sir Miles has truly an amazing memory, which he dismisses casually with the remark, "It's easy to remember something if you're interested in it." This memory has served him well on a number of other occasions. In 1919, for example, covering the Paris show for his editor, he was in the nightmare position of a correspondent whose copy and notes have been lost in the post. He was not unduly dismayed. Instead of bemoaning his bad luck, he set to and rewrote the whole report, including much minute detail about the new models – and all from memory.

Another time, when Miles Thomas was still a motoring journalist, he heard that a new light Rolls-Royce car was going to be introduced. He knew that Sir Henry Royce had a house near Wittering in Sussex, and so he went down for a weekend on the off-chance that he might see something to his advantage;

and outside a local hotel he saw a smaller Rolls than any being made at the time. Clearly, it was the new model. He had no camera with him, so he dashed into a chemist's shop, bought a Kodak and a spool of film, and then, as though he were a tourist taking snaps of the place for his album, he took several pictures of the new Rolls. They appeared later as exclusive photographs for his magazine.

Of course, to a man like this, as to Leonard Lord, success would have come in almost any career, but the fact was that it came through Morris; he was the first to recognize their worth. He was the catalyst and liberated their ambitions.

Early in 1926, a man came to see W R M in his office unannounced. This was the way it was in the early days; you had something to sell, an idea, a car, a factory, and so you went in and saw the chief, and you either did business or you didn't. They were good days, and Morris enjoyed them.

In this case it was a factory that was up for sale, a factory that made carburettors. Two brothers, Mr G H and Mr T Carl Skinner, had designed a carburettor of novel design. Their father had financed them, and indeed their carburettor was so good and ingenious that the prototype is now in the Science Museum at South Kensington. As they were working together, they called their joint enterprise "Skinner Union", or SU for short. Their invention had one fault, however; it was expensive to produce.

The SU was competing against such other carburettors as the Solex, the Zenith and the White and Poppe; their total output was small, and they were making a heavy trading loss. Naturally enough, their father did not want to continue financing such a seemingly disastrous venture indefinitely, and one of the brothers had withdrawn from the business. The other, Carl Skinner, had met Morris on a number of occasions and he felt it might be worth while trying to interest him in the venture. At most he could only lose his rail fare to Oxford, which was a few shillings; and he might conceivably redeem the failing fortunes of his company.

"Would you care to buy the SU concern?" he asked Morris, point blank.

"Why do you want to sell it? What's wrong with it?" asked W R M, countering one question with two.

"Here's why," said Skinner, taking the balance sheets out of his briefcase. "We're not making enough money."

Morris looked at the papers; the figures were in red ink.

"Hmm. A business that's losing money at this rate isn't worth a lot."

"Agreed, but it's a first-rate carburettor – why, you know it is. You've got one on your own private car. Now if we could fit SU carburettors on *all* the cars you made, then we'd be all right."

W R M considered the point, narrowing his eyes against the smoke from his cigarette, seeing not a run-down company for sale, but a vision of the future.

"How much do you want?" he asked.

Skinner named his price.

Morris set still for a minute or two, fingertips pressed together, thinking it over.

"All right," he said. "We'll have a contract drawn up in a day or two. Come back and we'll sign it."

"Thanks, W R M," said Skinner enthusiastically, shaking hands. He turned to go.

"Oh, by the way, what are your own plans now?" asked Morris. Skinner paused in the doorway, one hand on the handle.

"I'll find a job soon enough, I expect."

"Possibly, but what's wrong with the one you've got? I know nothing about making carburettors. I make cars. Your business is no good without you. Why not stay on and manage it?"

"What shall I pay myself, though?"

"Well, plough the profits back all you can. For the first few months just take out as much as you need for your own day-to-day expenses. I leave the exact figure to you."

So it was agreed.

Some time afterwards, Skinner suggested that new plant was needed.

"All right, go ahead and buy it then," said W R M casually.

Carl Skinner set about modernizing the plant. The total cost would be around £17,000, and so once more he arrived in the office at Cowley with all the relevant details, showing what he was buying, and for how much. W R M was not interested; in fact, he was almost angry.

"I don't want to see all this," he said, banging the papers on his desk with the back of his hand. "That's *your* job. What do you think I'm paying you for if I'm to do your work? Spend what you need. My secretary will watch the accounts..."

So Skinner bought his new equipment and out of that casual meeting the SU Company rose to great heights. Now their carburettors are fitted in many makes of cars besides those of the Morris group, and their output has increased from a few hundreds a week to many thousands. In the war, they made all the carburettors for the RAF Hurricanes and Spitfires up to the time of the Battle of Britain, in their small factory in Birmingham. And they also made carburettors for other aeroplanes besides; for Rolls-Royce engines, for the Vulture and for the Peregrine.

Then a patent for a fuel injector that two of the designers had taken out just before the war came into its own on the Mosquito fighter-bomber; and SU pumps, arranged in groups, were used on tank-landing craft at the D-day invasion.

In the very early days of the company, before Morris bought it, the carburettor had in its mechanism a set of leather bellows. Some critic said that while the basic design was good, he believed the leather would crack after prolonged use. The Skinners silenced him with the reply that as their family was connected with Lilley and Skinner, the famous shoe shops, the matter of good leather could safely be left in their hands...

Later, when W R M took over, someone asked, "What's the name of 'Skinner Union' going to be now – 'Morris Union'?"

Morris grinned.

"No," he said smiling. "People might think that I was trying to set up a rival to the Oxford Union..."

Shortly after he had bought up the SU Company, Morris went to the Motor Show, then held at Olympia, and as he stepped into the lift, there next to him was W H M Burgess, who had been associated with him before the war, and who was now in business on his own.

The two men shook hands.

"How are things, Willy?" asked W R M.

"Oh, so-so. Business isn't what it might be," replied Mr Burgess, for trade was not too bright anywhere in Britain during that year of the General Strike.

"No, I suppose it isn't," Morris agreed. Then, on an impulse, he said: "Look, I've just bought up the SU carburettor firm. We'll want a distributor to handle all the carburettors, and for service, and so forth. That's a task for you. Come and see me at Cowley after the Show and we'll fix it up."

Burgess went, and from that casual meeting another fine business, W H M Burgess Ltd., sprang up which is now under the direction of his son, Mr T M Burgess...

Morris believed that there should be no authority without responsibility, and vice versa. It was a belief that never let him down.

Not all W R M's enterprises were successful, however; sometimes he overreached himself. In the early nineteen-twenties, for instance, he bought the factory at Le Mans which had been owned by Leon Bollée, and determined to produce cars for the French market there – as Ford was producing cars for Britain in his own factory in Manchester.

The enterprise failed wretchedly. Morris had quite underestimated the great loyalty the French have for their own indigenous products. They were content with the cars made by Citroen and Renault, and Morris had to shut the works and bring his staff back to Cowley. When M Citroen established an assembly plant at Slough, though, the venture succeeded, and thousands of British Citroens are now on the roads.

In 1927, Morris bought the Wolseley Company, which was in a bad financial condition, and this was his most ambitious venture to that date.

The company had begun originally as the Wolseley Sheep Shearing Machine Company, and Herbert Austin, who later started the Austin Car Company, had been a designer there and had worked as a salesman for their products in Australia. He had indeed been responsible for the design and production of the first Wolseley car, a three-wheeler powered by a horizontal engine of two horsepower. Gordon Stewart, who became sole distributor for Morris cars, had also served with the firm for a time.

During the First World War they had produced a very fine car, and one that was made entirely by hand. Their profits were considerable: £137,000 was made in 1913, the year they introduced a quality small car called the Stellite, and when the war came this high profit increased considerably. They manufactured lorries, ambulances, munitions, and even aeroplanes in the First World War, and, as with most engineering firms of the day, the good times seemed to be endless. That they were not, was shown very sharply after the war. In 1919 the company acquired a site opposite the Ritz Hotel in London, and there they built, at a cost of £252,944, a magnificent building called Wolseley House, and set up showrooms and a sales organization.

Their profits that year, mainly from Government contracts which were still running, amounted to slightly less than half the cost of all this.

Then, with no Government contracts, and in the hard world of 1920, they registered a loss of £83,000. This was not particularly serious, considering the size of the company and its resources, for they were backed to some extent by the huge Vickers combine. But next year, in 1921, the loss had increased to £97,000; by 1923, to £327,000; and by 1924 to £364,000. In 1925 Wolseley's managed to cut this back to £189,000, but by then the battle was all against them; they had neither the strength nor the heart to fight back and win.

Thus it came about that, on a cold raw January day in 1927, the Bankruptcy Buildings in Carey Street heard the Senior Official Receiver run through the main reasons for the failure of this fine old company.

He blamed "Various causes...the moulders' strike, general trade depression, labour troubles in 1922, heavy writing down of stock in 1923, reduction of prices of cars to meet competition, and further writing down of stock, and in 1925 severe competition and reduction of stock values..."

Well, there it was. A dismal tale to hear on a dismal day, and one that gave no heart either to the employees of the company, or the shareholders, or to the public at large. There was probably another reason why the company had failed, although the Receiver did not mention it specifically: the fact that the company had abandoned one set of pre-war customers and could not find another.

In the days before 1914 they were known as the producers of a fairly high-priced quality car. People who bought a Wolseley paid a lot for it, but they bought a hand-made car, and one that, like the Daimler and the Rolls, also had a considerable amount of prestige value. If you ran a Wolseley, you must have means; it was the car for successful professional men.

After the war, they tried to make and sell more cars more cheaply; to utilize the methods of mass production. Their 10 hp cars sold, in 1924, at basic prices of £250 for the two-seater, and £285 for the four-seater, and the de luxe versions at £325 and £330.

It had similarly been suggested to Rolls-Royce, after the war, that they should abandon their policy of making relatively small numbers of expensive cars, and instead make more cars to sell more cheaply. This advice they disregarded absolutely. They knew that they had a market among moneyed people who did not mind paying more for a car that only a few could afford. Therein, indeed, lay much of the attraction of a Rolls. They were so expensive that they had a scarcity value. To own one was an outward sign of wealth and prosperity. Wolseley Motors, however, changed course and sank in uncharted seas.

On the last day of 1926 there was some excitement and speculation about the outcome of bids made for the firm. What was described as "one of the most interesting financial duels of recent years" began with an offer by "Mr Julius Turner, the thirty-year-old London financier", for the purchase of the company. It was believed that the Chrysler Company, of America, were backing him. Then it became known that Sir Herbert Austin was also interested as a potential buyer, although he made it clear that he was not asking on behalf of his company, but as a private individual.

At the Scottish Motor Show that year, someone asked W R M whether he would make an offer.

"I just want half an hour's quiet thinking over the matter before making up my mind," he replied. Before Christmas he had decided to bid for the company, and on 10 February 1927, the High Court announced that his offer of £730,000 had been accepted.

Ford used to say that he had bought the Lincoln car concern, which produced a superb machine, but uneconomically and at a price very few could afford, "more for personal reasons than because we wanted it". Probably Morris was actuated by the same motives; he disliked the idea that Wolseley might be bought by American interests, and neither was he anxious to see his British rival, Sir Herbert Austin, acquire the company.

So he bought it himself, and at once announced a thorough reorganization of the Wolseley works. The concern would be run as he ran Morris Motors, but still retaining its identity, and remaining entirely distinct from his other undertakings. It certainly needed reorganizing, for in the five years from 1920 to 1925 it had lost a total of more than £1,000,000.

Morris also stated that he would concentrate his efforts on the production of one car, the new six-cylinder Wolseley, that had been announced shortly before the business went up for sale.

Here he showed his unerring knack of going at once to the root of the trouble. This six-cylinder Wolseley was an expensive car, selling in the saloon version at £495, and as a tourer for £450, the sort of car Wolseley's had made well for years. He believed that, properly exploited, this car could do much to redeem their fortunes, and so it turned out.

The engine was tremendously strong, with seven crankshaft bearings, aluminium pistons, duralumin connecting rods and an overhead camshaft. It drove through a four-speed gearbox, and there were six brakes. *The Times* motoring correspondent of the day was impressed. "Wolseley designs have until recently left much to be desired," he wrote, "but the 16–45 horsepower two litre six cylinder chassis, although still capable of improvement here and there, is an unusually progressive and valuable production..."

So William Morris considered it, and indeed, that description might be applied to include the whole firm, for Wolseleys owned some of the finest manufacturing plant in the country. Many experts held that their presses for moulding mudguards and body panels were unequalled anywhere in Europe.

Thus Morris found himself possessed of a useful trinity to start with: a good car, a modern plant and a fine reputation. It was a trinity that became immensely successful. Each year the Wolseley models were improved, and the financial status of the factory improved with them. By 1930 they were producing no less than thirty-three different types, an astonishing variety when put against the two or three that many car factories produce today, and the two models that Wolseley now manufacture.

Then, "to meet the growing demand for distinction in luxury and equipment", the County series was introduced. These were luxurious cars, with silk-covered cushions for the heads of the passengers in the rear seats; parcel nets, smokers' and ladies' companions, and furniture hide upholstery. From the driver's viewpoint they were also interesting. They had radiator shutters that were controlled from the dashboard, an electrical petrol gauge, two spare wheels and dual dipping headlights.

In 1931 Morris paid a visit to the Wolseley works in Birmingham to give a final decision on their 1932 programme. First of all, he was shown an orthodox small quality car they proposed to market. He studied it carefully before he nodded his approval.

"That looks all right to me," he said briefly. "Go ahead."

"I have something else to show you," began the designer tentatively, and pulled a dust cover off an entirely new model: a car of modest size that was as roomy as many others of much higher price and horsepower, and possessed of an equal performance. The designers had moved the engine and gearbox forward in the frame so that part of the engine was now over the front axle instead of lying well behind it. This left more room for the driver and passengers.

W R M was impressed by the simplicity and ingenuity of the idea.

"Scrap the other," he said. "Go ahead with this." Thus was born the famous Wolseley Hornet series...

Wolseley engineers, of course, have always been ingenious. As long ago as 1934 they were incorporating an automatic clutch and free wheel in some of their models, innovations which America is now bringing out as "new". Then the "Nine" of the same year had what was called "pedomatic starting"; the engine started as soon as the accelerator pedal was pressed, another idea that has been hailed as a "new" feature on more recent American cars.

Every Wolseley model has been a car of high quality and first-class craftsmanship, and the same is true of all models of the Riley, another long-established firm that Nuffield bought, eleven years after he acquired Wolseley's. In neither case did he detract anything from the individuality of the cars; he preserved this as if it were his own.

The Riley Company have a magnificent record for fast cars. Their slogan is "As old as the industry, as modern as the hour", and it is true, and indeed an understatement, for there was a Riley factory in Coventry making equipment for weaving, which was the main industry there, long before the city became one of the capitals of the motor industry. Then, in 1896, the year of the Emancipation Run to Brighton, William Riley bought the Bonnick cycle business and began to manufacture bicycles, and the name was changed to the Riley Cycle Company. Two years after the foundation of the company, Percy Riley designed and built his first car, and until the First World War they produced motorcycles, cars and three-wheelers. They made the first engine to fit mechanically operated inlet valves. Before this, inlet valves were opened by "suction" on the down-stroke of the piston.

Always the Riley family were receptive to new ideas, and when the original founder handed on the business to his sons, their efforts made the family name famous. In the early twenties, the Riley Redwing was a car that became very popular; then the famous Nine, a sporty little saloon with wire wheels and a built-in luggage boot. In the early days of this design the overhead valve engine developed 35 bhp. Later, in a tuned form, it could produce 90 bhp, a very fine output indeed for such a small engine. Then the Imp and the Sprite...the Kestrel...the Monaco... the Adelphi, and the Eight-Ninety, an eight-cylinder Riley.

All were good cars, superbly built. The old Coventry craftsmen, who had been with the firm all their working lives, were proud of them and looked down on other mass-production "tin-ware".

Riley cars won the Ulster TT three years running, and in the expert hands of Freddie Dixon, the Riley was for a time almost unbeatable. The early ERA racing cars were also based on them.

Like the SU carburettor, in fact, the Rileys had only one fault: they cost too much. Enthusiasts coveted them but not all could afford them, and so, through the thirties, their financial position grew more serious. At one time there was talk that they would merge with the Triumph Company, whose own models, the Gloria, the Vitesse and the Dolomite, were of much the same type as the Riley. Nothing came of the scheme, however, and in September 1938 Lord Nuffield bought the Riley concern for £143,000, and then transferred all the shares to Morris Motors Ltd. for £100.

There were other acquisitions that he made, too...wartime developments, interests in other businesses... Lord Nuffield may never have had the chance of becoming a doctor of medicine, but as a man who could doctor a dying company, or as the *accoucheur* during the birth of a new one, this century has seen few to equal him.

CHAPTER FIVE

A Car for £100

*Most of all, I look forward to the time when it becomes a recognized
thing for a British workman to have his own car.*

Statement in 1926

In September 1927, Judge Crawford, sitting at Brentford County
Court, made a chance remark that aroused more controversy
than he could have imagined possible. In the course of saying
that, in his opinion, the great bulk of people in Britain were
living beyond their means, he added, as an afterthought: "Even
a County Court judge has no right to buy a motor car unless he
has private means, for his salary is not sufficient for that
purpose..."

At this time the stipend of such a judge was £1,500, and this
casual utterance became a national talking point.

"Is a man with an income of less than £1,500 a year justified
in running a car?" was the question of the hour. As Morris
Motors were then producing about seventy thousand cars a year,
and since they had made a profit in 1926 of more than
£1,000,000, their opinion was sought.

A director said: "It is one of our claims that the low-priced
motor car is within reach of any £400 a year man. There are
thousands of motorists with much smaller salaries than £1,500.
With the present low prices of cars, and the cheap petrol, it
should be possible for a married man with one or perhaps two
children to motor in comfort and safety on very much less.

Including depreciation, and all other outgoings, the £140 car ought to be run for twopence-halfpenny a mile…"

So successful had been W R M's production and business methods that the price of his cheapest car, the two-seater Morris-Cowley, had been brought down to £142 10s. Now he was considering the production of a car that would sell complete, and ready for the road, at £100.

This had been the aim of many other manufacturers, too, for years. Indeed, it was the ultimate at which many of them aimed, for they realized the magic of the description "The £100 Car". Thus, among public and motor trade alike, one of the questions of the day was: "Who will be first with it?" The question was, for a long time, largely rhetorical, but by the late nineteen-twenties, it seemed that either Morris or Austin might provide the answer, and interest grew with each new model that these two firms announced, in the hope that it might be the long-awaited, long-hoped-for car for £100.

Many believed that Austins would be the first to produce it, for they were making a great name with their "Baby" Seven, a car that has probably made more friends and given more pleasure than any other. Sir Herbert Austin, later Lord Austin, had designed the first one himself, so it was said, on his billiard table, and it was a success right from the time it was first marketed, in 1923.

Others favoured the Morris stable, however, and such had been W R M's astonishing success over the previous few years that he was believed capable of anything. He certainly had enough money behind him for any venture he chose to make, for in 1926 he had made Morris Motors into a public company, calling it Morris Motors (1926) Ltd., and floating it with the immense capital of £5,000,000, divided into 3,000,000 Cumulative Preference Shares and 2,000,000 Ordinary Shares of £1 each. The average annual profits of his firm for the previous three years had been more than £1,000,000 a year, and among its assets were National War Bonds amounting to £769,500, which had accumulated from undistributed profits. Thus,

Morris could afford to finance any new car he wanted to make. Indeed, he had so much that, if he wanted, he could produce a model and sell it at a loss, subsidizing this from his profits on the other cars in his range, and setting the financial loss against the more valuable advertisement and good-will that would accrue from a bold and popular move.

In the same year he did away with the rounded "bull-nose" radiator that had been as much his trade mark as his talisman since the first Morris of them all. So many people were sorry to see this old friend go out of production that Morris issued a personal statement explaining why he had decided to change it for a new design.

"We have adopted this new radiator so that we can obtain a more fashionable body line," he said. "I decided that it was not practical to fit the new longer and roomier bodies we are standardizing next year behind the old familiar radiator. Since a change had to be made, I have made it as definite and sweeping as possible."

As with everything he did, there could be no compromise, no half measures; off with the old and on with the new. To the general public, rightly or wrongly, this change was held to be vaguely symbolic of other changes, and rumours began to be put about that Morris was shortly going to produce a new small car of a type and size he had not so far made. The rumours persisted without either confirmation or denial until May of 1928, for W R M and Miles Thomas were too shrewd to interfere with this immense amount of free publicity. Then Morris announced that he had his plans ready for a new baby car.

"It has been apparent to me for some time," he said, "that there is a growing demand for the miniature type of car, and the new model has been designed to meet this demand. My wife was my sole confidante when I set out on this new venture. Then, as time went on, others had to be taken into our confidence, and eventually rumours were circulating that I was producing a new car.

"Agents began placing orders three weeks ago, and although we wished to keep it a secret we did not, of course, deny that the new model was ready…"

No price was given for the newcomer, for it had still to be calculated, but a spokesman for the Morris company admitted later that, "The new model will be two-thirds the size of this year's standard model. It will be narrower, and therefore much more easily garaged." For price, also, "a two-thirds scale can be taken as being near the mark".

The standard Morris-Cowley of that year cost £142 10s so people already began to expect this magic thing, "The £100 Car". Sir Herbert Austin was asked his opinion. He said that he "deprecated any suggestion of a fight between the Morris Company and the Austin Company", because the companies were on very good terms, and Mr Morris and he were personal friends. Their firms, then as now, were between them manufacturing more than half the cars produced in Great Britain, and their outputs were increasing every year. Sir Herbert thought it politic to add, however, almost as an afterthought, that "The Austin firm has been building the seven horsepower car for five years. They have sold nearly 100,000 of them, and naturally they have learned a very great deal of what was necessary to make a successful small car…"

Morris did not at first announce the name of his new small car. The fact was, he did not know what to call it himself. He referred to it once or twice as "the Morris Seven", but added, when people took him up on this name, that, "It is ridiculous to say that my intention in making the car is to compete with the Austin Seven. Some people seem to think that my idea is to try to push the Austin Seven off the market, which is absurd…"

The fact was that Austin's only serious competitor had announced that they were ready to enter the baby-car field, and whether anyone took this as a declaration of war or not was up to the individual concerned. Most people welcomed the newcomer, for where there is competition, then the price usually drops, and a better and cheaper article is made available to the buyer.

In the following month, Morris announced his prices: £125 for the touring car in either brown or blue cellulose, and £135 for the fabric saloon which was sold in the same colours. For £2 extra, the tourer could be fitted with a windscreen of safety glass; and for £6 10s all the windows of the saloon could be similarly equipped.

First descriptions of the new Morris were tantalizingly vague, and purposely so. Thus, one motoring writer said that it was "a smart looking youngster with attractive lines". Someone else who saw it noted, "It has the same shaped radiator as the standard Morris, but there is a different look to it." A third gave up the struggle: "The nearest description I can think of is that it looks like a miniature Rolls-Royce..."

This advance publicity was like the roll of drums that precedes a music-hall act: an essential part of the promotion. When W R M judged that the time had come for action, he announced its name: the Morris Minor, and he liked the alliteration so much that when he put another, larger car on the market in 1930, he called that the Morris Major.

In the late nineteen-thirties, the 8 hp Minor lost its name and was rechristened the Morris Eight, but in the late forties, with a redesigned body and chassis, the original name was brought back. As Lord Nuffield grew older, he liked to use the old familiar names, the Morris-Oxford and the Morris Minor. They were friendly names. They recalled earlier struggles and victories and meant a lot to him.

Extraordinary precautions had been taken in the few months preceding the announcement of the first Minor, to make sure that no unauthorized eyes should see the little car when it was being tested on the roads, and before it had been perfected. The first test car was built in a special shop away from the main body of the factory, and because its radiator was really a scaled-down replica of the new flat radiators on its bigger Morris brothers, no one was very keen to take it out locally for its first trial runs, in case it should be recognized. It was fitted with a dummy radiator shell quite unlike the familiar Morris shape, and then it was loaded on to a lorry at night, and taken miles away to be run on deserted roads in the early hours of the morning. The little car performed well; it had a stout-hearted engine, and just how good the design was can be gauged by the fact that hundreds of the original Minors are still running on the roads, and its engine and chassis were later the basis of the first MG Midget, which was destined to become one of the fastest small cars in the world.

But that is looking ahead; in 1928, the present was exciting enough.

The engine had overhead valves operated by an overhead camshaft, the drive for which was ingenious. The dynamo was mounted vertically in front of the cylinder block, and gears on the armature shaft meshed at the bottom with other gears on the crankshaft and at the top with gears on the end of the camshaft. Thus weight and extra moving parts were saved. This layout was later abandoned in the Minor cars, when they were fitted with side valve engines, but it remained as standard practice for MG Midgets until 1935.

The capacity of the engine was 847 cc, against the 747 of the Baby Austin, and ignition was by coil and battery instead of by magneto, which was still the more usual British method. The Minors had four-wheel brakes operated by foot pedal, while the hand-brake worked on the transmission, just behind the gearbox. This latter brake could be alarming if it was suddenly and fiercely applied by someone unfamiliar with the car, for wisps of smoke would seep up through the floor-boards from the burning brake lining. This fifth brake was popular with timid drivers. As a boy in Scotland, I remember being told by a maiden aunt to be ready with the hand-brake while we cautiously descended a mountain pass.

The cars were remarkably well equipped for their price and year. They all had bumpers at the front and back. They had ammeters, which some cars costing ten times as much do not specify today; a speedometer, a dashlight, an oil gauge, and on the radiator cap, the familiar temperature gauge of the time, that looked like a half-hunter watch.

The front two bucket seats, as on the present-day two-door Minor, tipped forward, in the words of the salesman, "to allow easy access to the rear". Just how easy this access was, of course, depended on how large and how old a person you were, but four people of average size could and did travel in the little Morris Minor in comfort. The windows did not wind up, but slid to and fro in their frames. This method of opening windows is still used today in some cars, notably the little Standard Eight.

The Minor was sold and serviced at fixed rates by the 1,750 Morris dealers in Great Britain – any of whom could also sell the buyer a small garage for his new car for an extra £10! There were many other light cars on the road at the same time, but the public took the Morris Minor to its heart, so that it became a classic among small cheap cars.

Though the Minor was indeed a fine little car, it was still not the £100 car, and for definite news of this, motorists had to wait until a few days before Christmas in 1930. The depression was at hand; the future seemed bleak and insecure. Three million people were unemployed; hungry, bewildered men were on the march from the depressed areas in Wales to tell London about their grievances. Works were closing every week, and people with money in the bank thought it more prudent to keep it there instead of buying a new car whose value would drop by a quarter as soon as they drove it out of the showroom and it became second-hand.

There were some very attractive hire purchase agreements available to tempt them. One firm even offered buyers a new car on payment of £5 down, and with no time limit for paying the rest. Even so, it was not impossible that car sales would become slower and slower until finally they almost stopped altogether. This had happened in 1921, and it could happen again in 1930.

Morris felt that this was the time for a bold announcement.

He took the decision on the last Saturday before Christmas, while he was out playing golf. A courier arrived from the works, and met him, as he walked off the links at Huntercombe, with a letter containing all the calculations he had asked for some time previously about costing a small car to sell at £100. They confirmed his own theory that a small two-seater could be sold for the magic sum, and that afternoon he sent word back to Cowley to go ahead with production. On the Monday, another line was added to the production lines already at his Cowley factory – one making the first £100 car.

There were those who doubted that he would succeed. The Clyno concern had burned its fingers very badly with their small car, which they had hoped eventually to sell at little more than £100. They announced the car in June 1928, and by the next summer they were out of business. Indeed a whole procession of small cars came and went in the late nineteen-twenties: the Bean, the Gwynne Eight, the Swift, and many others, but somehow the public did not take to them.

When W R M announced this latest venture, he explained that he had been prompted "because I believe that it is particularly in time of industrial difficulty that vigorous efforts to improve trade generally should be made. The introduction of this new model will give employment to a large number of British workmen, and will provide comfortable and reliable motoring for many members of the public…"

The Morris £100 two-seater was a brave attempt to meet a need, and much thought had gone into its design. A side valve engine of the same size as the overhead cam engine in the Minor saloon was fitted, and Lord Nuffield admits that it took nearly a year of trials and research to combine in the design of this new engine the degree of reliability and efficiency that he felt necessary. The car had been conceived for those who liked to look after their cars themselves – a tactful way of describing those who might not have the money to pay others to do so for them – and all mechanical complexities and accessories that might go wrong were reduced to the minimum. Yet, for its price and time, the £100 two-seater was well equipped, with adjustable friction shock absorbers, a three-speed gearbox, four-wheel brakes, a safety-glass windscreen, a driving mirror, a special bag for the hood when it was folded down, detachable sidescreens, and a spare wheel that fitted in the tail of the little car.

Morris said that it could do "upwards of forty miles an hour" and "over forty miles to the gallon", and he used one himself for a short while. "I like driving small cars," he said, and he still does.

Henry Ford used to offer his early cars in any colour the buyer wanted – so long as it was black. Morris produced this new car in one colour, grey. His earliest cars had been painted "Morris grey", and it is still one of the standard colours for new Minors. The upholstery was red, but the radiator shell, the windscreen frame, the lamps and some other parts that were usually chromium plated, were black. This innovation had two things to commend it: cheapness, and the fact that paint would not deteriorate like chromium if the car were left parked outside for long periods in all weathers. One of the selling points of the £100 car – apart from its price – was that it could be kept "anywhere". Suggestions were that it could be garaged in a tiny hut or in a tent, or just left at the roadside, for it was intended to appeal to people who had not previously been able to afford a new car of any sort, and who would probably not have garages attached to their houses.

In 1929, the Society of Motor Manufacturers and Traders had calculated in a review of the industry, that men with incomes of between £450 and £2,000 a year represented potential buyers of cars of one sort or another. In 1930, when the new two-seater was announced, statistics showed that 193,307 people in Britain had incomes between £400 and £500, but that there were more than twice this number who earned between £300 and £400. The Morris Minors, saloons, tourers and two-seaters, were aimed right at this great public: at men earning between £6 and £8 a week, but the demand for a two-seater car is not large in this country unless it is a sports car, so after rather less than one year, the two-seater £100 car was withdrawn. When a £100 car was again introduced in England, in the mid-thirties, it came from Ford of Dagenham, and was a four-seater saloon, aptly titled the "Popular" Ford – another title that has recently been revived for a new model.

The success of W R M's £100 car would have been much more marked had it been a four-seater, for a journey of any length made with two adults and any but the smallest child in the two-seater Minor was not a very comfortable undertaking. However, the price was right. A kerb weight of 12 hundredweight, or 1,344 pounds, worked out at 1s 8d per lb of car – cheaper than the price of English beef or mutton at the time! And the little car was also very cheap to run. Two people could drive from London to Brighton and back for three shillings; or go from London to Aberystwyth, a distance the guidebooks give as 210$^{1}/_{2}$ miles, for six shillings' worth of petrol – a feat that also demonstrates the size of the great gulf that exists between the cost of living then and now.

CHAPTER SIX

Safety Fast!

I have often tried to read books, but always, when I came to interesting facts or situations, I feel like putting the book down and investigating the truth for myself.

Especially do I feel this urge with travel books – no sooner do I read about a place than I want to go there at once myself. I always want first-hand and never second-hand information.

I like to find out for myself by actual experience under what strain metal will break, or at what temperature it will melt...

Sir William Morris, in 1929

Shortly after the First World War, Mr L F Pratt, one of the directors of Hollick and Pratt, the coach-builders at Coventry who were making the bodies for all the Morris cars, designed and produced an entirely new sports body for W R M to have on his own private car.

It was made from thin strips of oak and shaped like a boat, with the prow over the rear wheels, and each alternate lath of wood was stained so that the result was striking and unusual, like a long thin "humbug" on wheels. The bonnet was polished aluminium, and aluminium discs covered the spokes of the wheels. All the bright parts were nickel-plated.

Morris liked this car immensely. He would drive it about Oxford wearing a teddy-bear coat and a trilby hat pulled down over his eyes. The two-seater shell was very light and so the Cowley engine gave a far better performance than when it had to pull a more conventional saloon on a touring body. The car was stark, of course, by the standards of today, but somehow motorists then either did not seem to notice the weather, or else they were a hardier breed than they are now, or perhaps it was just that they wrapped up more thoroughly. Anyhow, in 1920 this special sports car was the last word in novelty, and many envious glances were cast at it.

Next year, in 1921, W R M entered three cars for the London to Land's End run, and all were awarded gold medals for their performance. One of them, Morris said later, was a new Sports Model, which would be produced "in a limited quantity", to sell at £395.

These sports cars were thus the first of this type to come from the Morris factory, but it was not for another two years that a sports car was marketed by them with a name of its own.

This, like much else with Morris at that time, was evolved – to borrow a phrase used of other thoroughbreds – "by Enthusiasm out of Investigation". The late Cecil Kimber was then general manager of Morris Garages, and a keen competition motorist. Into a stock Morris-Oxford chassis he fitted an overhead valve Hotchkiss engine, and instead of the heavy and rather clumsy body the car would have been blessed with if it had been a standard model, he knocked up a very thin sports shell with a swept-up scuttle – a design MG Midgets have to this day. He fitted two light bucket seats, staggered, in racing car practice of the time, so that the passenger's seat was slightly behind the driver's, which meant that the width of the body could be kept down.

Then he fitted light, cycle-type wings, fixed the hand-brake outside the cockpit, and decided to dispense altogether with a windscreen, both to save weight and to reduce resistance to the wind. As things turned out, this was not greatly missed, for the airstream was deflected by the scuttle and most of it passed over the top of the driver's head.

This car ran well; so well, indeed, that Cecil Kimber, who later became chief of the MG Car Company, entered it for the 1925 Land's End trial, in which it was awarded a Gold Medal.

Its top speed thirty years ago, powered by what was basically an ordinary 14 horsepower touring engine, was rather more than eighty miles an hour, an astonishing speed in those days, and nothing to be ashamed of even in these, when there are still some so-called sports cars with speedometers marked up to 100 that would be hard pressed to better it.

Clearly, in 1923 such a car had a future, but just what a future probably no one connected with its birth had any idea at all. Descendants of that first car have since been exported from the Morris Garages to almost every country in the world, and have won races wherever there are races to win, and they are largely responsible for the present wave of popularity enjoyed by British sports cars in America.

Everyone who knew anything about this first MG car in 1923 became enthusiastic about its performance, and many enquiries were received asking for replicas. There was obviously a lot of business to be done in this direction, and so a new company was formed to cope with it, and to make sports cars and nothing else.

The name chosen for this new venture was "The MG Car Company", and so many people were puzzled by these initials that in 1929, when the company moved to a new factory in Abingdon, they had to issue an official explanation: "Out of compliment to Sir William R Morris, Bt," they said, "we named our production the MG Sports, the letters being the initials of his original business undertaking, The Morris Garages, from which has sprung the vast group of separate enterprises, including the MG Car Company..."

For the next few years they produced sports cars very like the first one, but they were expensive to build, and in a world that was heading for an economic depression there would never be such a wide market for them as for something cheaper but still with a worthwhile performance. Motorcyclists who gave up two wheels, for instance, perhaps because of health or marriage, or the arrival of children, were faced with no sports car to which they could afford to "graduate". There were, of course, the "baby" cars of the day, but none of these were notably exciting for men used to travelling fast in safety, and the high-class sports cars of the late nineteen-twenties, such as the Bentley, the Lagonda and the Sunbeam, were expensive to buy and run, and beyond the means of most young enthusiasts.

W R M had made his fortune by producing an attractive touring car for the man of moderate means, and he believed that a considerable market also existed for an inexpensive sports car. It seemed to him that eventually others would realize this, too, so he determined to be first. Then he would indeed be The Ubiquitous Morris – a maker of all types of vehicles – touring cars, saloons, lorries, vans and sports cars.

The new sports car had to be cheap but with character; fast for its size, and yet reliable. Naturally, the price would be kept down if some of the parts that were being used in other cars of the Morris range could be used in the newcomer as well. Like the first Morris-Oxford, indeed, the new small sports car would be an in-between car; between the fast motorcycle and the fast large car. This was a tall order, but one that was brilliantly carried out. Under the direction of Cecil Kimber, the designers produced a small sports car that came to be the classic of all such: the MG Midget.

The first MG Midget was built on a strengthened Morris Minor chassis in 1929, and was powered by the ordinary overhead camshaft Minor engine of 847 cc with a redesigned exhaust manifold: the bore and stroke were 57 mm x 83 mm. All subsequent ohc MGs, from 1929 to 1936, had this same bore, apart from the PB, which had a bore of 60 mm. The name MG, in the octagon that has since become so famous, and which Cecil Kimber had first sketched out in his office at their first premises in Queen Street, Oxford, was stamped on the crankcase and also on the exhaust manifold.

In subsequent models this "Kimber octagon", as it came to be known, appeared in the most unlikely places. At one time, the sidelights were of octagonal shape, octagons on top of the aluminium camshaft cover concealed breathing holes, and the bolts that held this cover firm had octagonal heads stamped "MG". The oil filler was octagonally shaped; so was the radiator cap, and the top of the dipstick. Octagonal plates surrounded the instruments on the dashboard...the choke and hand throttle of the P types had octagonal plastic knobs, and also the gear lever... The octagon was everywhere.

The first MG Midgets had plywood bodies covered with fabric and tapering to a point at the tail. Vertical louvres were cut in metal strips that ran under the doors from the bonnet to the rear mudguards, to conceal the brake rods and the bottom of the battery which was fitted beneath the floor – and these louvres also gave the little car a sporty look. Driver and passenger sat behind a raked windscreen made up of two panels of safety glass with a central, vertical chromium rib to strengthen it, an arrangement that has reappeared since the last war, in a modified form, in the Jaguar XK 120 two-seater.

There was a three-speed gearbox in the early Midgets, and the gear lever had the end bent up so that it would never be far from the driver's hand. With an engine of this size, the sporting driver would have to make considerable use of his gearbox if he wanted the peak performance from the car. The universal joint was still the fabric disc used in the Morris Minors of the time, yet it stood up well to its harder task in the Midget, and the cars were even raced with these original fabric joints, which says much for their strength and reliability. This Midget, which was called the "M" type, had cable operated brakes, with ribbed drums, sensitive and light steering, and biggish André friction shock-absorbers that were adjustable.

It was a success right from the start, and so it has remained ever since. There have been faster sports cars, and better ones, but no other small and inexpensive sports car can have given so many people so much pleasure as the MG Midget. If any small boy were asked, between 1930 and the outbreak of the war, what sort of car he would like as soon as he was old enough to drive, his answer would almost certainly be, "An MG". The initials stand officially for "Morris Garages", but they also represent, like some algebraic symbol, an immeasurable amount of pleasure and happiness given. MG Midgets occupy such a unique position in the affections of sporting motorists for a variety of reasons; not least their size and saucy appearance, and also because for many enthusiasts they are their first motoring loves, and thus no car that comes after can ever equal them.

There are, of course, more concrete reasons for their popularity. They have always been reliable and there is about them something that encourages hard driving. Like a thoroughbred horse they respond to sympathetic and knowledgeable handling; and they have character.

The first overhead camshaft engines were compact, and strong with large bearings. The camshaft drive, by two sets of bevel gears through the armature of the dynamo that was mounted vertically in front of the engine, as in the original Minor, was so efficient that it was used in the engine of the 200 mph. "Magic Midget", with which Colonel "Goldie" Gardner has broken so many records both before and since the last war.

The power-to-weight ratio of all the Midgets has been good. The little Morris Minor engine developed around 20 bhp without any special tuning, and even with lamps and spare wheel, the weight of the first Midget was under 10 cwt. The present MG Midget, the TF, weighs $17^1/_2$ cwt, and its engine develops 56 bhp. The power output has kept pace with the increase of weight that has come with the modified design.

While work on the design of the first Midgets was proceeding, the MG technicians were also making out-and-out racing cars. It was the policy of the Company to incorporate the lessons they learned with their racing cars in the ordinary production models, a very sensible way of maintaining the breed.

The success of a team of Midgets in winning the Team Prize in the 1930 JCC Double Twelve meeting was the start of their interest and success in this direction, and their racing record is proud indeed. Their hey-day was from 1930 to 1935, when the Company withdrew from its racing activities, and in these five years MG cars were almost unbeatable.

They took the first six places and Team Prize in the same JCC meeting in 1931, the year when an MG was also the first 750 cc car to achieve 100 mph, and Lord Nuffield took a personal interest in Captain G E T Eyston's efforts next year to be the first man to drive an MG car with an engine of 750 cc capacity at more than 100 mph. This car was basically an "M" type Midget with a special cylinder block and pistons to bring the engine size within the 750 cc class. A cowl was built up behind the head of the driver and the passenger seat was covered in, so that wind resistance would be reduced. The wheels were "knock-on" Rudge-Whitworths; the standard "M" types had wire wheels secured by three bolts.

Brooklands Track was closed for repairs, so the car was tested on a straight stretch of road near Newmarket in the very early morning when there was no other traffic about, and then shipped to France for the trial run at Montlhéry. There was urgency in the air, for Sir Malcolm Campbell was known to be in Daytona for an attempt on the World's Speed Record, and he had a small Austin car with him, too, in which he hoped to be the first man to travel at 100 mph behind an engine of 750 cc.

At Montlhéry, the MG *entourage* heard with some alarm that he had broken the World Speed Record at 264 mph and then had driven the little Austin at 94 mph. Captain Eyston pushed the MG up to 97 mph, but they were both still below the magic figure, and it seemed that the reason might be that the MG engine was running at a lower temperature than it liked. Mechanics knocked up a crude cowling over the radiator and another attempt was made. This was successful, 103.13 mph over five kilometres being achieved, an astonishing speed for such a tiny engine.

In 1932, a Midget was the first 750 cc car to achieve 120 mph, and the number of "firsts" they won became almost monotonous as the years went on... First and Second in the 1,100 cc Class, and the Team Prize in the 1933 Mille Miglia...holder of all Class "H" International Records. Then, in 1934, more victories; all the Class "H" records and also all the Class "G" International Records... First, Isle of Man Mannin Beg Race... First, RAC

Ulster TT... The BRDC Empire Trophy Race... First...American AC Grand Prix. Then, in 1935, more Firsts – in France, England, Ireland, Australia, Germany... Everywhere they appeared, they added honour to their name.

Their speed and roadholding, and their extraordinary reliability, became the yardstick by which the small sports and racing cars of the time were judged. Just before the last war, Colonel "Goldie" Gardner drove his streamlined "Magic Midget" at more than 200 mph in Germany. The highest speed recorded was 206 mph, but for such record attempts the mean speed of two runs, one going, one coming, is taken.

No one in the early days of the MG concern could possibly imagine these ultimate results from their interest in making a Morris-Oxford more potent, but enthusiasm begot success. In the background Lord Nuffield, who was never greatly interested in racing his cars – although he had been keen on racing his cycles in the early days – would smile indulgently at the ambitions of his colleagues to make the MG internationally famous. Thus, when Lord Nuffield sold the assets of the Company to Morris Motors, Ltd., in 1935, and it was announced that there would be no more racing, the news was received with sorrow by thousands of enthusiasts. So intense was the feeling, indeed, that the Company had to issue a special statement that gave their reasons. It was published in the *Sports Car* magazine for August, 1935:

> Lord Nuffield has said there are to be no more MG racing cars. This announcement came as a shock to all and sundry connected with or interested in the sport, for since 1931 one or other of the highly successful racing types has either won outright, or most certainly been well in the picture, in almost every British and Continental event for which the cars have been eligible.
>
> In fact, if the upholding of British prestige abroad can be laid at the door of any one *marque* that surely is MG...
>
> The real reasons for discontinuing the building of racing machines are as follows:

First of all, the Directors have decided that, at all events for the present time, racing for the purpose of development has,...served its useful purpose.

Another reason, rather more obscure, purely concerns racing itself and has no bearing on the commercial aspect, namely, that we are handicapped out of British racing, through no real fault of the handicapper. It is simply a case of carrying a fundamentally unsatisfactory system to its logical conclusion.

It stands to reason that a car which very frequently wins must inevitably have its handicapped speeds increased to a greater amount than the "also rans", whose development and speed capabilities are to the handicapper far more nebulous.

This attitude can better be understood when it is appreciated that MG racing cars are securing first place in almost all the Continental events in which they compete and which are run on a class basis without handicap systems...

Lord Nuffield had repeated the experiences of his cycle-making days when he had raced his cycles to get them talked about; and he had ceased when his object was successfully attained. Thus there passed from the active racing scene some of the finest small racing cars that have ever been built. MG cars, of course, still win races all over the world, and trials and rallies for which they are entered by private individuals without the official backing of the works. The MG Car Company still produces models that have followed the famous ancestors of the early thirties, incorporating many virtues first learned on the race-track: road-holding, rigidity of frame, positive steering and worthwhile acceleration. And all live up proudly to the slogan that the firm first made famous more than twenty years ago: Safety Fast!

CHAPTER SEVEN

£26,000,000 Given Away

The man who would give money is compelled to do a great deal of hard thinking. Is his gift going to do harm or good? Money has tremendous power and can do either. The responsibility of the would-be giver is great. If he is a decent man, he cannot escape it.

I find a similar difficulty in lending money. Is it going to help the borrower on, or land him in a worse position next week? I have come to this conclusion: the man who would pay back never borrows; while the man who would borrow never pays back...

W R M in 1927

All through the early years after the First World War, Morris and his wife stayed on in the Manor House at Cowley, next to the works. Eventually, the noise and general expansion of the place forced them to move, and because W R M was a keen golfer and had often played on the course at Huntercombe, near Henley, they decided to move there and live in part of the clubhouse. Then changes threatened in the organization of the club and it seemed likely that they would have to leave. Reluctantly, W R M made his most ambitious purchase on his own account, and bought up the entire golf course, clubhouse and all. Thereafter, he circularised some of the old members of the club and asked them to play on his own private course. Nowadays, while he does not play a great deal of golf himself, he often follows friends round on summer evenings, his wife's Scots terrier trotting along contentedly at his heels.

Their home was formerly a gardener's cottage, modest and quite unpretentious. He has never owned a racehorse or a yacht, and he has been quite happy without them. His happiest moments apart from his work have been spent golfing, or, nowadays, mending the clocks and watches of his friends.

Henry Ford used to walk through his factories at Detroit, now and then throwing aphorisms to those who accompanied him. One of his favourites was, "Money is like an arm or a leg – use it or lose it." Morris believed that the best use it could be put to was in helping others. The total value of his gifts now exceeds £26,000,000, and this sum does not include many donations of up to £10,000 that he has given away to help people and good causes. The largest single gift is the Morris Motors stock, valued at £10,000,000, which he gave as a charitable trust in 1943 to form the Nuffield Foundation, a gift that was hailed as "the largest and most notable in the history of the nation". The most unusual is probably that of "twelve pairs of football boots, twelve jerseys, twelve pairs of shorts and as many pairs of stockings", which he had given five years previously to the Balmoral Thistle Football Club, a club that drew its players from one of the poorer districts of Aberdeen.

The secretary was most surprised to receive it. "I took a chance," he said delightedly. "I read that Lord Nuffield had given money for sports purposes in England, so, well, I told him of our club's financial worries and at once he offered to give us what we wanted..."

The Nuffield Foundation was made, Lord Nuffield explained, "from resources which have been built up through private enterprise, in the essential importance of which I am a firm believer", and it had several objects. First, he wanted money to be spent on medical research and teaching; then, on the organization and development of medical and health services, and on scientific research and teaching in the interests of trade and industry; and, finally, on the pursuit of social studies and problems connected with old age.

The Nuffield Foundation has helped many scientific undertakings. It made a direct grant towards the research into penicillin by Sir Howard Florey, when Sir Farquhar Buzzard, Regius Professor of Medicine at Oxford at the time, told Lord Nuffield that unless more money could be found for the purpose, the work could not go on. W R M is keenly interested in the future application of this drug, and is proud that its application for medical purposes was evolved in Oxford and with his backing.

At one time, in the late nineteen-thirties, Lord Nuffield was giving money away at the rate of £8 a minute: wealth that grew from an idea, translated into riches first by himself, then by a few others, and then by hundreds and thousands, all making the cars of which he had dreamed.

The Romans had a saying, "He gives twice who gives quickly", and few benefactors have ever given so spontaneously as Lord Nuffield.

In November of 1936, for instance, having given the immense sum of £1,250,000 to Oxford University for medical research, a special Convocation was called in the Sheldonian Theatre there so that official thanks might be paid him. All the University dons and doctors, wearing the glorious robes of their full academic dress, met to honour him in this historic building, and to fix the University seal to the trust deed of the gift. The Chancellor of the University, Lord Halifax, was there, sitting on a dais in his gold and black robes, and to his left and right the heads of the colleges and halls, each in his or her gown and hood. Lord Nuffield sat in a comparatively inconspicuous place, wearing the scarlet robes of the Doctor of Civil Law, a degree which had been conferred on him some time previously by Oxford in token of his earlier kindness. Lady Nuffield watched the ceremony from a seat in the body of the hall.

Speech after speech, in English and Latin, paid tribute to him for his munificence, and he sat on through them, embarrassed at the praise, wiping his hands with his handkerchief. Eventually the speeches came to an end, but instead of the organ playing solemn music, as is usually the case on such an important occasion, Lord Nuffield stood up, nervously fingering the edge of his gown.

He bowed towards Lord Halifax, and began to speak. His voice sounded thin and hollow in the huge vastness of the circular theatre, and people craned forward to hear him better.

"I rise by special leave of this House, to thank you most sincerely for your kindly appreciation of my endeavours," he said. "But I rise in particular for the reason that I understood through my secretary this morning that the money already subscribed was not sufficient to produce the effect I anticipated."

He paused. No one coughed. No one moved. No one seemed even to breathe. Outside, cars hooted and newsvendors called the racing results, but within the Sheldonian there was only room for silence. Lord Nuffield folded his hands in front of him and went on speaking.

"I understand that the amount necessary is not one but *two* million pounds, and while I have been sitting here, I have been thinking it is a great pity that such a scheme should not be brought to fruition as soon as possible. I would not like to think, after all the work that has been put into the scheme, that it should fall through.

"I would like, therefore, to increase my donation of £1,250,000, which has already been accepted, to two million pounds. I would hate to leave this building with the feeling that the scheme was not complete."

For a moment the silence remained, an emptiness of wonder, and then a roar of applause swept through that old building, the like of which it has not heard before or since. Bearded and white-haired dons were on their feet cheering like schoolboys whose house has won a rugger match. Some shook hands, grinning; others stamped their feet; many wept. It was a fantastic scene of regard for a man and for an announcement of generosity that was without parallel in the University. The cheering continued for five minutes, and through it all Lord Nuffield, who had taken his seat again, sat on quietly, but with a pleased smile.

Usually at Convocation the Chancellor does not give an address, but this was no usual Convocation. Lord Halifax broke the rule to speak of "our feeling of unspeakable gratitude to Lord Nuffield for the princely munificence with which he has completed a gift which, in the words of the decree, was already 'unexampled'. He will forgive me if I say no more because, when things are done as he has done this great thing this afternoon, words are not easily found that fit the deed. Therefore, with his permission, and with yours, I shall formally now close our proceedings…"

The afternoon was over; but the giving was not.

At about this time, Nuffield was receiving more than five hundred letters a day, each one marked "Personal" and "Private", and each one asking for money. Sometimes they amused him; sometimes they irritated him. He would say ruefully, "If I opened all the letters sent to me thus, I'd spend the whole of my working days doing nothing else."

No amount of such letters, though, could stem the astonishing flood of generosity for worthwhile causes. He gave anywhere, in the most unlikely places. He passed through Colombo, on a business trip to Australia, and heard that a school for deaf and dumb children was in difficulties. At once he gave the governors 35,000 rupees, about £2,625. In New Zealand, he went to a luncheon where the talk turned to helping crippled children. Someone mentioned that he had given up his own house in the country so that the children could have somewhere to stay. The gesture so moved Lord Nuffield that he at once offered to endow it with £50,000.

Another time, at a public dinner, he sat next to Sir Wemyss Grant-Wilson, at that time honorary director of the Borstal Institutions, and the talk finally swung round to their work and aims.

"Our real trouble," said Sir Wemyss, "is to get the parents to visit the boys. They live so far away and they cannot all afford to come. What we need is a fleet of motor cars to carry them."

Casually Lord Nuffield – then Sir William Morris – asked how much this would mean in money.

"About £10,000," replied Sir Wemyss, making conversation.

"I'll give it to you," said W R M, making history...

By 1936 he had given away so much, and there were so many confused accounts of the amounts and the causes he had helped, that Morris Motors Ltd. had to issue a statement to newspaper editors so that they could know the true state of affairs. The total at that time was about £7,000,000. By June 1939 the company sent another circular: the total was now £12,000,000. Today, it is more than twice this sum.

He has given money to almost every conceivable good cause, but always the needs of medicine and surgery have been closest to his heart. He has given millions to help research and to found new hospitals and medical schools and to help others that were already doing good work. Then he provided £1,036,000 for the site, the building and the endowment of Nuffield College. Originally, he wanted this to be solely an engineering college, but he was persuaded to allow the articles of the association to be more liberal. One of the aims of the college is "the practical study of social problems".

Although he has been generous to Universities, Lord Nuffield has never been convinced that a degree is necessary to success. In engineering, he knows that it certainly is not.

"If I had a son," he says (and it is one of his deepest disappointments that he has no family), "I would send him to a public school, and if he were going into any of the professions a university degree would be essential. But not if he were going to be a production engineer. The best place for practical experience is in the engineering shops... As I see things, the only way for him to get his qualifications is through hard experience in business and the shops. Much as I think of a university

95

education, I cannot see how that can be obtained in a university…"

Millions more of his money went to form a special trust to help the distressed areas, for Lord Nuffield has created sixteen trusts and many other endowments.

"My interest in medical matters is now one of the busiest sides of my life," he says. "Since I've retired I seem to be busier than ever."

It has often been said that Lord Nuffield can well afford to be generous, "because if he doesn't give the money away, then it would only be taken from him in Income Tax". At the present rate of Income Tax and surtax, which leaves only thirty-six men in Britain with incomes of more than £5,000 a year after taxation, Lord Nuffield pays enormous sums to the Inland Revenue; but the money he has given away has been money spent out of capital – which would not have been taxed at all. There has been no need for him to be so generous; yet, some years before the last war, he became so annoyed at this campaign of sneering at his generosity that he seriously considered making no more gifts at all. "They need only keep on a little longer with it," he would say, "and I'll stop giving money away…" But he never has carried out his threat, although the criticism has not entirely ceased.

In the nineteen-thirties, Lord Nuffield was one of the pioneers of the iron lung. These are used in the treatment of infantile paralysis. The patient, who cannot breathe normally, is placed inside an airtight box with his head and neck sticking out of the end through a sponge-lined collar. Bellows withdraw the air from the box, and the patient's lungs expand; then air is pumped back again into the box and its pressure forces the patient to breathe out again, and so the cycle goes on. People have been kept alive by such lungs for many months; babies have even been born in them.

Lord Nuffield's "iron" lungs were not made of iron at all, but were actually plywood boxes, designed by an Australian, Edward Thomas Both, who had the idea after hearing an appeal on the radio for such a machine to save the life of a sick man miles from the nearest hospital. Professor R R Macintosh, Nuffield Professor of Anaesthetics at Oxford, made a film that showed the working of an iron lung, and asked Lord Nuffield to see it. He was impressed, and went back to his factory in Cowley determined to make these "lungs" himself.

"That film convinced me that no hospital in the Empire which needed an iron lung should be without one," he told his friends afterwards. "Motorcar factories can produce almost anything. I decided to use the brains at my factory and have them collaborate with the medical men. We will make improvements as they are discovered..."

Each lung cost him £98, and by the time the war stopped their production, because it became impossible to get the little motors that powered them, he had made and given away three thousand. He did not think twice about the money he was spending. "All I want to do is to save life," he said. "It seems a dreadful state of affairs that children are dying because hospital authorities cannot get hold of an iron lung in time."

This lonely, withdrawn man without any children of his own has saved the lives of thousands of other men's children. Yet very few of the people he has helped ever write and thank him. When Lord Nuffield is asked about this sort of ingratitude, however, he shrugs his shoulders. "I don't do it for the thanks," he says curtly, "but because it needs doing. Others could do the same..."

Nevertheless, he is delighted when he receives a letter from a father or a mother, saying that their child has been helped by one of Lord Nuffield's gifts. He keeps such letters for days, and answers each one personally. Yet he is always embarrassed when people mention his gifts to him. "What I have done, others can do," he says. "If they don't, they will regret it when they die. I just want to pass out feeling I've done my best for mankind..."

Lord Nuffield was still plain W R Morris when he began giving his money away in 1926. He had returned from a business trip to the Argentine, where he was surprised to find that few British businessmen out there spoke Spanish well. He believed that if Britons could talk business in the language of the country, they might do much more business, and so he gave £10,000 to Oxford University to endow a chair of Spanish studies.

Such unexpected generosity, and from a man of Oxford city, surprised the University. The records were searched, and it was announced that this was the first gift of such magnitude that any Oxford businessman had ever given to the University. It might have been the first; it was millions short of the last.

Twenty-seven years later, in 1953, when Lord Nuffield was being made an honorary fellow of the Faculty of Anaesthetists of the Royal College of Surgeons, he described how, equally casually, he decided to endow another University seat – a Chair of Anaesthetics at Oxford – the first to be established anywhere.

Lord Nuffield explained that he had been given several bad anaesthetics, and only one good one – by Professor Macintosh.

"The time came when I suggested, after long experience of anaesthetics, that something should be done," he said. "I suggested a Chair of Anaesthetics at Oxford. Some said: 'Why give the money for that? Any doctor can administer an anaesthetic.' I said, in the words of the American, 'Says you.'

"There was still some uncertainty at Oxford. I said, 'Well, if you don't *want* the Chair here, I can take it to London.' They accepted it…"

In 1927, just twelve months after he had surprised Oxford by his first gift of £10,000, W R M surprised St Thomas' Hospital, London, by making them a gift of £104,000. The hospital secretary was astounded. "It is the biggest donation, apart from legacies, that we have ever received," he exclaimed. There was a deficit between the hospital's income and their annual expenditure of nearly £20,000, and Mr George Roberts, a governor and one of the almoners, who knew Morris, had asked him whether he could help at all. W R M was glad to do so, but was surprised at the publicity he received as a result. He had intended to say nothing at all about the gift, and indeed it only came to light because the hospital authorities were too excited and pleased about it to keep it a secret.

This was the first really big donation; others began to follow it quickly. In April of that year, W R M was in Birmingham at the opening of a new Morris service depot. He took the opportunity of giving £25,000 to help local hospitals. By June, he had said he would give another £15,000 to hospitals in Coventry, if a carnival they were arranging was successful in raising £500. Then he promised the Radcliffe Infirmary at Oxford £38,000 to build a new maternity home; and in October he announced that if an appeal for £120,000 for extensions reached £80,000 he would double his original offer...

The ancient Greeks believed in a cycle of events; victory for one side meant defeat for the other, and this, in turn, led to resentment and eventually to victory again. W R M's generosity started a train of events of an entirely different kind. He gave his money away with a remarkable spontaneity, and, as a result, his gifts were discussed all over the country. Many people decided to show their approval of his actions by buying his cars, and so the more he gave away, the more money he made; an astonishing chain reaction that was only stopped by the Second World War.

On the last day of March 1930 Sir William Morris, as he then was, told some friends that he was going to take a 10 per cent tax free dividend from his firm. By not taking any dividend at all before, a reserve sum of £2,000,000 had been built up, and so his dividend amounted to £200,000.

"That sum has been allowed because I have not enough to pay the charitable institutions who are looking to me," he explained almost apologetically, because always, publicly and privately, he had preached the virtues of ploughing back profits into the business, so that plant could be renewed, overheads reduced and even more money made.

He has also been generous to those who work for him. Like Henry Ford, he has no love of trades unions, but for all that there has never been a serious strike at any of his factories. Nearly twenty-five years ago he announced a welfare scheme whereby should any of the eight thousand employees on his payroll die their dependants would receive between £50 and £100.

Then in 1941, at the time of the great bombing raids on the Midlands, when he had thirty thousand working in his factories, he announced that the next of kin of any worker killed by enemy action would receive a similar sum.

He was one of the first big British employers to begin pension schemes for his workers, and in 1936 he made them all shareholders by presenting them with stock worth more than £2,000,000, to create a trust fund for them.

Lord Nuffield knew their minds, for, as he would often say, "When I go through the shops I know exactly what the workers think of me, and the conditions under which I ask them to work... I try to treat them as I would expect to be treated if I were in their shoes."

W R M always decides himself who and what will benefit from his gifts; he never takes the advice of others on the subject.

"When I've to sit round a board table and listen to what the majority decides, I'm finished. My best decisions are made in a minute," he says.

Just how much money has he left now that he has given away more than £26,000,000? The question has often been asked, and some kind of calculation can be made.

In 1936, when Lord Nuffield put on the market a considerable part of the share capital of Morris Motors, which he sold for something like £3,750,000, the shares he still held were valued at about £16,000,000. Then he gave £2,125,000 worth of stock units to his employees, and so, if this gift is added to the 1939 value of Morris shares – 32s as against the 36s of 1936 – he was at this date worth a possible £12,000,000.

Today he is worth only a fraction of this sum.

In 1943 the Nuffield Foundation cost him £10,000,000, and on top of that he gave away thousands and thousands of pounds for Services welfare during the war and just before. For instance, in 1938 he wrote to the Secretary of State for War:

I have been greatly impressed and encouraged by the wonderful response to the National Appeal for Voluntary recruiting... Not less remarkable has been the willing acceptance by all classes and parties of the principle of universal service, calling for equal sacrifices by all...

I am anxious to make some personal contribution towards the comfort and well-being of those who are giving up, however temporarily, the ordinary course of their civil occupations and their home surroundings in the service of our country.

For this purpose I intend to place in the hands of Trustees one million shares in Morris Motors, of a present value of approximately £1,500,000, yielding today an income of some £105,000 per annum, to be devoted towards improving the facilities for recreation and enjoyment of the Militia, Territorials and other Forces...

I intend this gift to be a permanent memorial to the spirit which animates us today.

Shortly after this, he was reading a magazine and he saw a paragraph about some European girls in Singapore who were organizing a charity cabaret. Their enthusiasm and initiative impressed him, so he cabled £500 to help them out.

Morris gave quickly and generously to any good cause that attracted his attention.

Thousands of Royal Air Force aircrew received five shillings a day as "leave money" from him. He also maintained suites in hotels in various seaside towns so that they could enjoy their leave at his expense. There is also the Nuffield Centre, in London, for the Services on leave, a legacy of the war that still remains.

In the early days of the war, too, someone told him that the troops with the BEF in France had no radio sets and thus no means of hearing the news or being entertained.

"We can do something there," he said – and gave £50,000 for troops' recreational facilities, of which he asked that £15,000 should be used to buy them radio sets.

It grieved him that the dependants of those who were killed in the RAF might be suffering hardship, so he gave £250,000 to the RAF Benevolent Fund "to relieve all forms of distress".

On Poppy Day, 1939, he dropped a cheque for £100,000 in the box of a flag-seller, and she was so surprised that, so it is said, she forgot to give him the poppy for his lapel!

The extent of his generosity is staggering, and still the good work goes on, and indeed it will never entirely cease, for even when Lord Nuffield is no longer here, people will benefit from his kindness and foresight.

"When I make money," he says now, "it enhances the security of the workers dependent on the organization I control. For the rest, the only pleasure that riches give me is that they enable me to attempt to buy happiness for others…"

Then the shrewdness of the man shows itself:

"But note," he adds dryly, "I do not give money to people for the happiness that *money* might bring to them. Good health brings far more happiness than money can ever do…"

Here he speaks out of his own experience, for although Lord Nuffield's own health has never been robust, by his endeavours and his mighty gifts for medical research he has helped improve both the health of his countrymen, and the heritage of those who will come after.

CHAPTER EIGHT

The Man with the Red Wig

Hours of work for me simply represent the maximum I can get out of the day.

W R M in 1929

Once, in the late nineteen-twenties, when the Morris factory was making more than £1,000,000 profit every year, W R M was prevailed upon to take a holiday, for although the Cowley works closed for two weeks every summer, he would still come into his office every day. He had worked so hard as a young man that he had no hobbies at all, and any free time dragged slowly and without enjoyment.

"Get right away from everything," one of his directors advised him. "You'll come back a new man. A rest will do you good."

"But I don't want to be done good to. I'm fit enough as I am," Morris protested, but eventually he was persuaded to go to the South of France for a long holiday.

The man who had persuaded him was thus astounded, only a day or two afterwards, to see his chief in one of the Cowley workshops, rolling himself a cigarette, his face creased against the smoke of the one already in his mouth, watching the men at work, chatting to them, missing nothing.

"But I thought you were in the South of France," he said, surprised.

Morris grinned a bit sheepishly, like a schoolboy caught in the pantry near a cake.

"Oh, I was, but I got bored stiff. I had to come back. Couldn't lie about there all day doing nothing..."

This life appealed very strongly to many other citizens, however, and the begging letters arrived in their thousands every month.

"I need to have six people sorting mail from 8.30 in the morning, and it's eleven before I receive my own private letters," says Lord Nuffield. "Sometimes, when I'm thinking of making a gift, I'm almost persuaded not to by fear of the after-effects. Since I started giving, my worries increased five hundred per cent..."

In the spring of 1938 a man, who scorned the crude approach of a begging letter, evolved a complicated plan to kidnap Lord Nuffield and hold him to ransom for £100,000. He called himself John Bruce Thornton, but his real name was Patrick Boyle Tuellman, and he had been born in Scotland, the son of a German watchmaker.

He liked life at its most gracious level, a preference that was at variance with his talents as a wage earner. He had been a telegraph messenger, an apprentice joiner, an electrician, but he had tastes and a longing for a life above anything he could afford in these jobs, and he was not too particular how he made his money, so long as he could live well in gay places, such as Cannes, Nice and Monte Carlo.

Some time before he thought out his plan to kidnap Lord Nuffield, Thornton met another man, Arthur Geoffrey Francis Ramsden, whom he had known after the First World War. As neither was in work and the weather was good, they used to drive in Ramsden's car through the country lanes near Henley.

One day Thornton nodded casually out of the car window.

"See that place over there?" he asked, jerking his head in the direction of a house set back from the road.

"Yes, I do," replied Ramsden. "Why?"

"It's Huntercombe, where Lord Nuffield lives."

"Oh, yes?"

"Yes," said Thornton again, turning to him with the air of one who has had a brilliant idea.

"Look, I've thought of a way to get rich quickly. Let's kidnap Lord Nuffield and hold him to ransom. I've watched him for a long time. I travelled with him on the boat when he went to Australia last, so I know quite a bit about him. He'd be an easy man to grab – there are no guards or anything of that sort, for nothing like that has ever been attempted in England before... You see...it'd be *bound* to succeed."

Kidnap; it was an ugly word on that sunny day. Ramsden did not like the sound of it at all. He steered the conversation into a less provocative channel.

A few days later, Thornton returned to the subject. He had given a good deal of thought to it, he said, and he'd worked out a plan that was absolutely foolproof. In brief, Thornton explained, he would find a way of reaching Lord Nuffield in his office, and once they were alone he would warn him that he was armed, and that if Lord Nuffield shouted for help or made a false move he would be shot.

Then they would walk out to a car which Ramsden would have outside with its engine running, and drive to some quiet cove where a small yacht would take the three of them out to sea. When they had Lord Nuffield alone, they would force him to make out a draft of money – £100,000 was the sum Thornton had in mind – and then blindfold him, gag his mouth with sticking plaster and bring him back to England. They would cash the draft, dump him in a field, and escape themselves.

Thornton had also thought of a way in to Lord Nuffield's office: he would pretend to be a journalist wanting to write an article about him.

They would call Lord Nuffield "Kelly", so that no one who overheard them would know whom they were discussing. Then Thornton said he would adopt the name of Dr Webb when they had "Kelly" aboard their yacht, and would persuade him to write two letters on Morris Motors paper, which they would acquire in some way beforehand. One letter would go to Lord

Nuffield's secretary, explaining his sudden absence by saying he had gone on holiday; and the second would serve as a letter of introduction for Thornton when he presented his cheque for £100,000 at the bank for payment.

There were other, more lugubrious details to which Thornton had given some thought. He "happened to have" a red wig; he would also peroxide his own hair and dye his eyebrows and moustache, and fit gold caps to his front teeth, so that he could ring the changes between red hair and blond. Then he had a set of corsets he could wear under his shirt to make him appear slimmer.

When Ramsden asked why he already possessed this astonishing collection of disguises, Thornton explained vaguely that he was divorced from his wife, who lived in the South of France where he owned some property. They were not on speaking terms, but he liked to go down there from time to time to examine his houses and lands, and when he went he always wore some disguise. He travelled as a stranger, he said, because his wife might not allow him to see his own possessions if she knew his identity.

This seemed an odd story to tell and have believed. In fact, the more Ramsden considered the whole idea, the more lunatic did it appear, and he wished he were well away and out of it. Thornton, however, had no such qualms; to him it all seemed clear and sensible.

"When I go with the letter of credit," he told Major Ramsden, as though the actual kidnapping had already been successfully accomplished, "you stay on the yacht and listen to the wireless. If you hear I've been arrested, you'll know that Nuffield has not played the game with us. Take him out to sea – and dump him."

It sounded melodramatic and simple, the way he spoke, commandingly, but the thought must have occurred to Ramsden that if *he* were left with Lord Nuffield aboard the yacht, and Thornton had the £100,000, then what was to prevent Thornton leaving him in this embarrassing position, and not returning either to help him or to share the profits?

Major Ramsden felt that no good could possibly come out of such a scheme, but he decided to keep Thornton's confidence in the hope that he might learn more of the plan.

This he did, and quickly.

Thornton's greatest problem was to find a boat where Lord Nuffield could be kept a prisoner, and also a place where this vessel could be moored without causing any local gossip, until they needed her. He discussed this question with Ramsden, who knew a small village on the East Coast called Pinmill, where boats were often for hire. A yacht could anchor there for weeks without causing any comment, and the place had the added advantage of being on a good road from Oxford. They drove up there, but no boats were available for hire, so they motored to another yachting centre near West Mersea, and hired an eleven-ton auxiliary cutter, *Pierette*. She had a cabin and an auxiliary engine, and they loaded her with provisions, and returned to Oxford.

A few days later, the two men went down to Colchester, where Thornton hired a portable typewriter on a £5 deposit, and then drove out into the country, where Thornton typed a letter to the man they hoped to kidnap; a letter he hoped would act as a passport to see his lordship:

"My dear Lord Nuffield," he began, "I am writing a series of articles on outstanding and prominent business-men in England and Europe for publication in America.

"From what I have heard of yourself and your work, I am extremely desirous of including you in this series as being the outstanding personality in England today.

"As the background of these articles is a personality and character sketch of an individual, I would be extremely grateful if you would grant me an interview to suit your own convenience.

"I am not interested in seeking financial assistance in any shape or form for myself or for any charitable purpose.

"A draft of the article itself will be submitted to you for your approval before publication.

"I am visiting business centres in England, and will be in the vicinity of Oxford during the week beginning May 23rd, and shall take the liberty of phoning you, asking when you can grant me an interview."

He signed the letter "R C Wilson", which, looking back, as he had time to do later, was not very wise, for he had signed for the typewriter in the name of "A G Wilson", and had given his address as 22 Sheep Street, Northampton. He marked the envelope "Personal", addressed it to Viscount Nuffield, Morris Motors Ltd., Cowley, and then they drove on to Chelmsford where he posted it. Then Thornton had second thoughts about the typewriter. No two typewriters have quite the same sort of letters, and he feared that this machine might be traced back to him, so he returned to the shop and bought the machine outright. The one loophole in his plot, so it seemed to him, was now closed up.

Unfortunately for him a far larger one was left unsealed. This was his companion Major Ramsden, who was by now thoroughly uneasy about the whole wild scheme, and wished he had never met Thornton or heard of his idea for making money quickly.

They spent the night at Thame, and next day Thornton telephoned Lord Nuffield as he had suggested, spoke to his secretary, and asked for an appointment.

Lord Nuffield's secretary consulted his chief's engagement book, and said that an appointment could be made at six o'clock on May the 24th. Would that be convenient?

"Most," replied Thornton thankfully, secretly surprised how well everything seemed to be going, and then hung up.

That evening he busied himself typing out a further letter. It had occurred to him that someone else might also be present when he met Lord Nuffield, who would have to be got out of the way before he could announce the real object of his visit. The problem was to persuade them to leave the room without arousing any suspicion, for obviously he could not order them out himself. The only thing was to persuade Lord Nuffield to do so, and thus he typed another letter which he planned to give Lord Nuffield with the words, "I have this letter of introduction to you", which would ensure that Nuffield at least glanced at the sheet, if only to see who had recommended his visitor.

When he read further, he would receive a shock, for the note began:

To Viscount Nuffield:
Read this carefully before passing any remark, and don't show it to any other person.

1. I am packing two automatic pistols of large calibre and will immediately shoot you through the guts if you attempt to raise alarm or suspicion. Any help will be too late to help you.

2. You will ask me to come and see your children's hospital.

3. Cancel any appointments which you may have for today.

4. Walk out with me to my car and the chauffeur will do the rest. Don't on your way out attempt to make a run for it; it means instant death to you and to anyone who attempts to interfere.

5. Smile and chat and be cheery even if it hurts. Don't raise suspicion anywhere, or…

6. I am a quick and accurate shot with a gun, but do exactly what you have been told and you have nothing to fear.

7. Place this letter on your desk with the remark, "The writer is a close personal friend of mine".

8. Dismiss anyone who may be in your office.

9. Don't attempt to leave me under any pretext whatever.

10. Offer me a cigarette before the car starts and light one yourself.

11. Make sure that we are not followed by anyone of your folk; it is fatal for you.

12. Jump to it; I am in a hurry.

That last afternoon took a long time to die. By five-thirty Thornton was ready for his part; disguised, with gold caps on his teeth, his letter inside his jacket pocket. False number plates were on their car. Ramsden was wearing a chauffeur's cap and a dark suit. In silence, their thoughts on what lay ahead, they drove towards Cowley.

They were nearly there when Thornton broke the silence. After such a long time his voice sounded thin and strange.

"We can't do it today," he said hoarsely. "I've burned the wig. Drive into the country and we'll change the number plates."

A candle had dropped on the wig, Thornton went on vaguely. It had been ruined. The scheme would have to be postponed.

Ramsden pointed out that he would have to telephone Lord Nuffield if he wanted another appointment with him. He would wonder what had happened; after all, the fellow was a millionaire. You couldn't mess about with a chap like him.

"I see that," agreed Thornton. "Ring him up and say that our car has broken down. Tell him anything you like."

Ramsden nodded, saying nothing; he might almost have been expecting this sudden change of plan. He drove up a street by the side of the gigantic Morris works, and stopped.

"Be back in a minute," he said, climbing out of the car. "There's a phone box here."

Thornton nodded.

"OK," he said, lighting a cigarette, "Don't be too long." Ramsden walked round the corner where he knew there was a telephone box, and dialled the number of Morris Motors.

He cleared his throat, looking over his shoulder in the direction of his parked car. His face was pale and damp with nervousness. He pressed button A.

"We're here," he said hoarsely, and described where he had left the car.

Within minutes, it was all over.

Black Wolseley police cars hemmed in their little car, and Thornton was led through the main gates of the Morris works and into the works office to be searched. The police found a shoulder harness strapped under his left arm, with a Browning automatic in it, empty. Another pistol was in a case inside the car, and ammunition to fit them both; also, the red wig in a box. It had not been burned at all; Thornton had just lost his nerve at the last moment. So far as he was concerned, the pity was that he had not lost his nerve much sooner, for Ramsden had been in touch with the police shortly after the first mention of kidnapping "Kelly". He was, in the crude language of the underworld, a "stool pigeon".

As Thornton stood there, half clothed, with police making an inventory of the things found in his pockets, Lord Nuffield came into the room, smiling.

"I'm the man you came to collect," he said.

"I don't know what you're talking about," retorted Thornton.

The police found an odd collection of goods inside the little Ford car: a bottle of peroxide, which Thornton might have used to dye his moustache; three hair sticks, a diary, a box of adhesive plaster, a set of stays and a knuckleduster. Thornton's story was that Ramsden had invented some device which could be fitted to the carburettor of a car to economize on petrol, and he had been promised a quarter of the profits if he could persuade Lord Nuffield to fix this contrivance on all the cars his firms produced. The note he had intended to show to Lord Nuffield before he kidnapped him was, so he explained, part of a film scenario he was writing.

This must have sounded pretty thin stuff that May evening in the hostile, excited atmosphere at Cowley; and it did not seem any more likely when the yacht *Pierette* was boarded by the police, for they found chains and padlocks, which could have been used to hold a prisoner still; rubber gloves, forceps, scalpels, a hypodermic syringe, and phials of hyoscine, of hydrobromide, digitalin and strychnine sulphate, with bandages and hypodermic needles. It was suggested that, should Lord Nuffield have proved stubborn and refused to write the notes, an operation would have been performed on him without anaesthetics to try and make him change his mind.

Thornton spent that night in a cell in Oxford police station, and next morning was charged with being in possession of firearms with intent to endanger life.

When this charge was read out to him at the time of his arrest he had protested, "Not with intent to endanger life – the things weren't loaded"; but that, as prosecuting counsel pointed out, was "no answer to the charge".

From his first appearance in Oxford Police Court Thornton was remanded. "I am in communication with the Director of Public Prosecutions," reported the Chief Constable.

When Thornton came up again a new charge was preferred: "That he, between April 1st and May 24th, did incite Arthur Geoffrey Francis Ramsden, to conspire with him to kidnap one Viscount Nuffield."

He was committed to Birmingham Assizes on the additional charge that, "On May 24th, at Oxford, he was in possession of two automatic pistols and ammunition with intent to endanger life."

Thornton pleaded "Not Guilty" to them both. He denied he had any responsibility, or the intention of taking part in any plot to kidnap anyone.

"I met Ramsden in Cannes," he affirmed. "His finances were bad, but he had a scheme for saving 40 per cent of petrol consumption in motor cars. He wanted to explain the idea to Lord Nuffield.

"To get hold of him, I would have to impersonate a doctor and call on the resident house surgeon at the Wingfield-Morris Orthopaedic Hospital and say I was a doctor interested in orthopaedic work. From seeing the surgeon in the hospital it would be easy to get in touch with the big man himself."

This whole idea, he said, had struck him as being "hopeless".

He went on: "Ramsden then put up the idea that I should pretend to be a visiting American journalist who wanted to include Lord Nuffield in a series of articles he was writing."

Mr Justice Wrottesley, trying the case, did not think much of either story; nor, indeed, of the whole shabby affair.

"It has been said that the story is of an astonishing nature – melodramatic to the last degree. Here are all the trappings of a detective novel – disguise, pistols, harness to carry pistols under the arm, red wig, false number plates on the car…

"It has been suggested that the whole thing is a fantastic story," he told the jury, summing up. "But of course there is great force in this: …all these extraordinary things *were* in Thornton's possession…"

"If Thornton's story is true," he continued, "all he ever knew was that Ramsden said he had invented some device for saving petrol… If that was all that was going to be done, why the two pistols, why the red wig, why buy a typewriter in a false name, why have false number plates?…"

The jury took two hours to make up their minds about Thornton's intentions; and their deliberations resulted in seven years' gaol for him.

And Major Ramsden – what happened to him?

Nothing at all. He was exonerated from any part in the business, but not before counsel for Thornton had said some bitter things to him.

"Being hard up," he asked, "did it ever enter your head that Lord Nuffield – generous as we know he is – would have dealt very generously with the man who saved him from a painful experience?"

"It never entered my head," said Major Ramsden indignantly.

"Did you hope for any reward as a result of this case?"

"I do not," replied Ramsden stoutly, "And I never expected one."

Which was just as well; for he never got one, either.

And Lord Nuffield, how did he take all this commotion? Very quietly indeed; quite in character. It might have been happening to him every day for years.

On the night of the attempted kidnapping, he knew what was planned, of course, and he was sitting in his office. In an adjoining room was a friend, Mr Kennerley Rumford, husband of the late Dame Clara Butt, the contralto. The works were empty, and, save for night-watchmen on their rounds, seemingly deserted. Here and there a few lights gleamed faintly through the vast glass windows of the assembly shops, but no shift was working.

Some men were still on the premises, though: the Morris Motors Works Band, practising. Their notes sounded eerie and disembodied in the huge, lonely vastness of the place.

As soon as Thornton had been taken away, Lord Nuffield walked into the rehearsal room, and the bandsmen lowered their instruments awkwardly, seeing the Chief unexpectedly in front of them.

"Well, boys, what do you think of it?" asked Lord Nuffield briskly. "Two men have just tried to kidnap me!"

The band were surprised to silence: no one had words to match the moment. Later, one of them remarked ruminatively, "The Old Man made a joke of it."

The fact was, the Old Man thought the whole thing rather funny.

CHAPTER NINE

War in Three Cities

Life is a practical job. Thinking about it may help some people, but the longer I live the more certain do I become that a man can only find himself in the work he does. I am afraid that this is not a fashionable doctrine...

W R M in 1928

The twenties and the thirties were the golden years in the life of Lord Nuffield. What went before was the struggle and the worry and the midnight working; the dread of failure and the dream of success. What came after was money and generosity, and honours; but none of them could sweeten the later years and make them as exciting as the early time had been. After those days, of course, anything would be an aftermath; for never have there been such years for one man, and nowadays England is too small for another man to show such high endeavour, and reap such tremendous returns.

In the late nineteen-twenties, then, and at the peak of his success in making vehicles to travel on land, W R M turned his attention to the air. Looking back now, when a jet plane can fly from London to Khartoum in less time than a fast car can be driven from London to Land's End, it seems incredible that only a few years ago Amy Johnson flew her tiny little aircraft from London to South Africa and broke the record, and received from Sir William Morris the gift of a four-seater MG, smartly turned out in black and red with a tiny mascot of Jason, her aeroplane, on the radiator cap.

Morris admired her pluck, but many people thought that this gift was not made in admiration of her flight, or because he liked a pretty face, but because he wanted her advice, for he was interested in manufacturing aero engines on a large scale – and who better to advise him about some aspects of this than such a pioneer of solo flight?

In those days, it was confidently expected that soon people would own and fly their own aeroplanes, much as they ran cars, without thinking anything of it. Cartoons appeared in the newspapers showing "jaywalkers of the skies", and policemen directing the air traffic with tiny engines and propellors strapped to their backs. Those who said that such a state of affairs would never come to pass were laughed at, and it was pointed out how others had said much the same thing when the motor car was in its infancy, and they had been proved wrong within a few years.

Sir William Morris shared the general enthusiasm for flight, and he intended that his firm should be ready to meet the demand he expected. He set up a factory to manufacture aero engines at Wolseley's, in Birmingham, and in 1932 his first 150 horsepower engine successfully powered a two-seater Hawker Tom Tit. When this news was known, Morris admitted the extent of his interest in the experiments.

"I have been considering the production of an aero engine for the last three or four years," he confessed. "There is no doubt that, as the aeroplane improves and becomes safer, so it will be more popular as a means of transport, and the more aeroplanes that are sold, the more reasonable their price will become. This also applies to the cost of the engines. I have no wish to build aeroplanes myself, and no intention of doing it, either. I am just out to sell engines to the manufacturers.

"There is no big demand for engines at the moment, but it was because we realized that the sale of aircraft would gradually climb – as did the sale of motor cars – that we decided it was time we set out to design engines for manufacturers. It might be true to say that the aeroplane is today where the motor car was in 1914. When aeroplanes are sold in sufficient quantities, there is no reason why they should not be as cheap as the light car of today..."

In theory, there is of course no reason at all; but in practice an aeroplane of even the most modest size costs far more to run than a light car. First, there must be a suitable hangar where it can be kept; and an airfield near the owner's home; then, a certificate of airworthiness is needed, and this must be renewed from time to time. Also, the maintenance of an aeroplane is expensive, and so are the dues that must frequently be paid when it lands at alien airports. All these add enormously to the running costs. Thus, it seems unlikely that the small private aeroplane will ever become a serious competitor of the small car in this country, though in Canada and America and Australia conditions are quite different.

Morris, like many others in the early nineteen-thirties, was wrong in his view, but he also had another aim when he opened the aero engine factory: he wanted a plant that would be capable of producing aeroplane engines in quantities, should any emergency arise and the RAF need them. Years later, when he disagreed with the Government of the day on the question of the best way of producing aeroplanes in large quantities, he described what had been in his mind. "I put up the aero engine factory because I realized that, in time of national emergency, firms with experience of building internal combustion engines might be called on for national defence, and I wished to play my part..."

Those words showed his vision, for no one could have imagined just what a worthwhile and vigorous part he was to play in the Second World War, when the Morris works at Cowley, under the official title of Civilian Repair Organization, salvaged nearly eighty thousand wrecked RAF planes, repaired them and sent them back again fit to fly and fight. How valuable this was in the early, chaotic days of the war can be gathered from the fact that at the end of the Battle of Britain, the RAF had only nine planes in reserve. But between July and December of 1940, the CRO had returned no less than 4,196 machines ready for battle, and each one renovated at about one third of the cost of a new aeroplane.

In the early thirties, though, when Morris opened his new factory, few thought there would be another war, and fewer still in authority believed there was any great need for aeroplane engines. The Air Ministry, seemingly, was quite satisfied with the number and types it had already, an outlook that Morris did not share. He knew how much it would cost him to make an aeroplane engine and sell it at a profit, for he had more experience in the art of engineering on a large scale and in flow production than most other manufacturers in this country, and he held that the Air Ministry was paying too high a price for its engines.

He mentioned this casually to a friend, and the remark was repeated and quoted until it reached the ears of some people at the Air Ministry.

"It got back to certain political interests there, where, apparently, it was not received with favour," says Lord Nuffield now with a wry grin. He never had a very high opinion of the talents and capabilities of those who chose to spend their lives in the cushioned safety of Government Departments; so it did not altogether surprise him that his opinions and ideas did not meet with their favour.

"Throughout the whole development of aero engine activities, my advice was characterized by a complete lack of support from these interests," he recalls, but adds in fairness, the technician warming to his kind, "The technical side gave freely such help as they were allowed to put forward…"

Progress was slow. Morris was not sure which were the best kind of engines to build, and the Air Ministry did not hurry to give him advice. It was five years from the time he first became interested in the mass production of aero engines before they would accept one of his engines for trial.

To try and draw attention to his new venture, he entered planes in the King's Cup Air Races of 1933 and 1934, powered by Wolseley engines, but official interest was still lacking. He sent the Minister descriptions of two engines he was making. Would they be of any use? Was he working along the right lines? Back came the reply: "We think it is improbable that the Air Ministry will be in a position to utilize either of the types of engine you describe in your letter..."

All right, thought Morris; but if they don't want those, why on earth can't they say what types they *do* want?

Other European countries were arming now, especially the Germans and the Italians. They were buying and building warplanes, fighters and bombers. Only in Britain was there a lack of purpose and policy. The British were aboard their little island, which had not been invaded since 1066. Surely they were safe in 1934? Anyway, there was always the Navy...

The mood of the people was against anything to do with war, but the need to be prepared was growing more obvious every day, although few in power and politics took heed of it, and those who did were unable to make the Government change its policy of drifting. Lord Nuffield kept on writing to the Air Ministry, asking them what kind of engines they could use that he could make. He wanted to help; all he lacked was direction, but still the polite official refusals of his assistance came back to him. He was completely baffled.

As late as 28 March 1935 – four years before the outbreak of war – the Air Ministry sent him a letter in which it was stated that "The Air Ministry has already five aero engine firms under contract, and, except for one, which also relies on civil work, it is a matter of the greatest difficulty to provide them with enough work to keep them employed. It is very unlikely, therefore, that anything can be done to consider any engine of your design for Royal Air Force purposes..."

If the Government were not interested, then no matter; Lord Nuffield was a rich man, and altogether he spent £500,000 of his own money to keep the accounts of his aero engine factory entirely separate from those of the other firms in which the public held shares. This aero engine factory he regarded as his own private venture, and he was determined to bear any losses himself. He believed that his engines were good, and that reactionary interests were withholding them from the Air Force. He made up his mind to break those interests.

In July of 1935 he wrote to the Air Minister of the time, Lord Swinton, and asked for an interview with him. There seemed no point in continuing further with the hitherto unsatisfactory basis of working.

"During the past year or so", he wrote, "the managing director of Wolseley Aero Engines, Ltd., has done all he can to interest the Air Ministry, but he has been unsuccessful. It is this point on which I should like to have your advice, as it occurs to me that if there is no prospect of our doing business it would perhaps be wiser to liquidate the company and abandon our aero engine activities altogether..."

Lord Swinton replied that it would be "quite impossible" for him to see Lord Nuffield on the day suggested. "I have two meetings, as well as a Cabinet, and two deputations," he wrote, excusing himself, "and I have to be in the House and to attend a public dinner, so that literally every minute of my time is taken up..."

Nuffield kept on pressing for information. A man who has sold cars is not usually easily turned aside; a man who first made the cars, and then sold them, is even more tenacious. Thus, eventually, a meeting was arranged between Lord Nuffield, Lord Swinton and Lord Weir, who was adviser to the Government on the air expansion programme.

W R M explained that he was willing to do anything within his power to help the country; that he would obtain the designs, and manufacture at his own expense, either an American type of engine that was available, or else the Bristol engine.

"I was not asking for any financial assistance from the Government, or doing anything that would be any burden on the taxpayer," he said afterwards, when he explained that his offer had not been accepted. "Finally, feeling that the Ministry might change its mind, I sent engineers over to the United States to investigate the latest production methods over there, so that we could add to our experience and knowledge."

The Government of the time, however, had their own ideas of engine production, and they did not coincide with those of his lordship. They wanted a number of car manufacturers to agree to set up "shadow factories" in various parts of the country. One factory would make one component, a second factory some other part, and so on, and then the engines would be assembled in a further factory.

One reason for this method of working was that no one factory would be expected to know all about the engine; they would simply concentrate on their particular part of it, and security would be easier to maintain. Against this, Lord Nuffield had a serious objection: a single bomb on the assembly plant could ruin the whole scheme, for the component parts that might still be made in other associated concerns could then never be assembled. Also, he considered that in precision engineering of the nature involved in aero engines, unity of control was essential.

For these two reasons, and also because he already had a factory capable of producing complete engines, he refused to join in with the Government shadow factory scheme. Instead, he offered to build two thousand Bristol engines at the same price as the Air Ministry was already paying. The offer was not accepted.

In desperation, the managing director of Wolseley's wrote to the Air Ministry: "Both Lord Nuffield and I are anxious to help the Air Ministry to increase its production capacity. We are an established aero engine company equipped with the very latest plant and inspection machinery which can be quickly duplicated, and we can commence deliveries of two or three Bristol engines per week in approximately six months' time.

"It is our considered opinion that this company, by rendering the service we have suggested, would be of greater value to the

Air Ministry than by co-operating in the shadow organization scheme…"

He received in reply a letter in which it was noted that "Whatever success may be achieved by your company in the production of engines during the expansion period, the scheme would contribute little towards the development of the largely extended capacity requested for an emergency, nor would there be created thereby any prospect of Air Ministry orders for engines from your company after the expansion was completed…"

Which official language, rendered down, meant one word: No.

Nuffield was by now thoroughly annoyed.

"I am asked to put up a shadow factory at Government expense," he told his friends, "when I already have an aero engine factory standing there doing nothing!

"If that's not waste of public money, then I don't know what is. It's incredible to me. I even offered to build parts in my factory. No, it must be an Air Ministry factory. Perhaps they imagine I can't *make* pistons," he added bitterly, pulling out his little cigarette machine and rolling himself a cigarette with fingers that trembled with rage and frustration. He was sure that he was right; it annoyed him intensely that the Air Ministry should also think that they were.

"It's staggering. Really, you can hardly conceive that such a state of things can exist. I don't understand it, and I don't suppose anyone else will, either."

Lord Nuffield's annoyances were still not over: but let him tell the story himself…

"On August the 27th, 1936, I felt that the time was ripe for me to make up my mind about this aero engine work generally. We were very busy doing considerable important Government work, and our higher executives spent much time in contact with the War Office. The design and production staffs of Morris Commercial Cars Ltd. were engaged in making specialized Army transport vehicles, for instance, among other things.

"On this day, anyhow, a conversation took place between our factory and the Air Ministry. We told them that we were closing down the aero engine plant and that a letter confirming this would be posted that night.

"Well, imagine our surprise when next morning we had a letter from the Air Ministry *asking for three hundred engines*!

"It's an astounding state of affairs. They must have said: 'They're going to close down – get that tender out as quickly as possible!' "

According to Lord Nuffield, this is what took place at his last meeting with Lord Swinton.

"I was turned down flat. I said to Viscount Swinton, 'Then I take it you do not want our engines?'

"He said, 'Well, no.'

"I said, 'Well, do you want us to make Pratt and Whitney engines?'

"He said, 'That is your business.'

"I tried again. I said, 'Is there not *anything* you would like me to do?'

"He replied, 'Well, I suppose it comes to that.' "

Lord Nuffield picked up his hat, and his gloves, and walked across the room. At the door he paused.

"God help you in time of war," he said – words that later became famous.

Although Lord Nuffield found the Air Ministry hard to deal with, his relations with the Army had always been very friendly. As long ago as 1925, Morris Commercial had produced a six-wheeler which gave good service to both the British and the Indian Army. Then they had pioneered the use of half-track vehicles that could pull heavy guns. Thus, within a week of his public announcement that he would not co-operate in the shadow factory scheme, Nuffield was given a much larger job: that of bringing up-to-date the mechanization of the Army, and the ground section of the Royal Air Force. This followed half an hour's talk he had with the Prime Minister, Mr Stanley Baldwin, who had stressed how anxious he was to have Nuffield's co-operation; and Lord Nuffield assured him that he was only too happy to co-operate in any workable scheme.

This was the result: that the Wolseley Aero Engine plant would manufacture tank engines, and that research units would be established there to work on new ideas for improving the mechanization and mobility of the Army.

This ending to the unhappy dispute seemed to please everyone. Lord Swinton, after shaking hands with Lord Nuffield, told the House of Lords, "I can give this assurance...that Lord Nuffield's great personal capacity, and the great organization which he has created, will be used to greater advantage in the service of the State."

Nuffield made one of his own rare visits to the House of Lords to hear him speak, and sat, face pale, head down, legs crossed, during the Minister's speech.

"The creation of this controversy is most regrettable," he agreed afterwards, "and it was not of my seeking. First of all, I want to say there had been no quarrel or dispute between me and the Air Ministry. The whole matter is simply a difference of opinion on principles of production. As the largest manufacturer of internal combustion engines in this country, I can claim *some* experience in this direction. The Air Ministry have their point of view, which, of course, they are entitled to. It just so happens that we do not see eye to eye..."

Thus, out of confusion and harsh words, the co-operation of the leading motor maker in the country was secured. It was a great pity that the incident received all the publicity it did, for Germany and other interested nations knew then what was planned for the Wolseley Works in Birmingham, should war come; and when the bombing started, Wolseley's were selected for special attention...

It was also at the Wolseley Works on Sunday, 3 September, 1939, shortly after Mr Chamberlain had made his broadcast to the nation announcing a state of war, that the directors and higher executives of the Nuffield Organization met to discuss what they would do.

The chairman of the meeting, the late Mr Oliver Boden, managing director of the company, left them with the words: "The manufacture of civilian vehicles will have to cease the moment war work becomes available... We must be prepared to do everything the country needs. Nothing else counts..."

Nor did anything else count throughout the war in the tremendous factories that Morris had built in the three cities of Oxford, Coventry and Birmingham. It was largely due to Lord Nuffield's foresight that his factories were able to turn over so quickly to the manufacture of war material. In 1936, for instance, a team of senior British officers, including Britain's foremost tank expert, General Sir Giffard Le Quesne Martel, visited Moscow to watch the autumn manoeuvres of the Red Army, and returned much impressed by the performance of Russia's light tanks, which, it seemed, had been based on a tank originally designed by an American engineer called Walter Christie. General Martel told Lord Nuffield about this, and he was so interested that at once he sent an engineer over to America to buy one of the original tanks on behalf of Morris Commercial Cars, Ltd.

An American Neutrality Act in force at the time prevented the tank being shipped as it was, so the turret was taken off and the body crated and stamped "Tractor". Then it was loaded and brought to England.

The secret of this tank lay in the ingenious suspension that Christie had designed: large wheels to carry the track, working on swinging arms as opposed to the small wheels favoured by the British War Office. Speeds of more than 40 mph were possible with a great improvement in manoeuvrability.

W R M was impressed, and, with the backing of General Martel, who was then Assistant Director of Mechanization at the War Office, he began to manufacture tanks. He formed a new company, Nuffield Mechanizations Ltd., in Birmingham, next to the Wolseley Works, and at one time in the war more than one quarter of all the tanks made in this country were made there. In 1940, indeed, this was the only factory making a complete tank in one place, and using W R M's old ideas of "flow-production" to do it.

Tanks and trucks, staff cars and ambulances by the thousand …millions of rounds of ammunition…eighty thousand aeroplanes put back in the skies…landing craft…submarine engines…all these occupied the men of the Nuffield Organization during the war. Other factories sprang up under the aegis of the parent concern at Cowley. Lord Nuffield, driving from one city to another, would stop his car suddenly and say to directors who were travelling with him: "This seems a likely site. Let's have a factory here" – and within weeks the foundations would be laid and new buildings arise.

He was in his sixties, and troubled by sciatica; often he would have to leave his office and walk across to the works medical centre for treatment. Still he worked on. Tiger Moths… torpedoes…engines for Lancasters…the Neptune, an amphibious vehicle that could carry a 17-pounder, or a 3-ton bomb… Bofors guns…all these engaged him. The strain was enormous, for he knew many secrets, and his was the brain that had conceived the whole organization and the energy that inspired it. Some time during the war, his hair turned from grey to near-white.

CHAPTER TEN

Interim Assessment

Without optimism, no one could live in this world...
Sir William Morris, 1930

Afterwards, nothing was quite the same. It was not only that everyone was tired, that cities lay in ruins and the towns still standing were tawdry and shabby. The feeling went deeper than any of these things. It was as though, in achieving the final blow of victory, Britain had strained her great spirit; the nation was depressed spiritually and, for a time, the zest was not left in living.

Lord Nuffield felt the change acutely. There was not the same feeling of victory that had come in 1918. There was little savour in this peace; there was overmuch fear, of the atom bomb, of traitors, of so many things and people alien to the English scene.

The daily drive from Huntercombe to Cowley seemed longer than it had been. The roads were more crowded, of course; perhaps that had something to do with it. You had to wait five minutes, sometimes, before you could cross the High in Oxford, the traffic was so heavy.

Also, there was so little left for W R M to achieve. He had attained all his ambitions, save the one he had wanted most of all: to be a surgeon. The car for £100...a good cheap touring car for people without a great deal of money...sports cars... limousines...tanks...vans...lorries. Millions made and millions given away; degrees, civic honours, the membership of learned societies. All these were his, and yet perversely, he sometimes thought that he had never wanted them very much.

"I wonder what I'd have done if I'd been born rich, and had a good education?" he asked a friend at this time.

"You'd have been one of the best surgeons in the world," came the instant reply.

"Ah, yes, I'd forgotten that. I'd have liked to have had the chance," and he would look wistful and a little sad. The courage that has served him so well in making great financial decisions would have helped him, perhaps, in the investigations of new surgical techniques.

Sometimes, too, in the evenings, when the dusk creeps quietly over the lawns of Huntercombe, and he is alone, perhaps listening to Gilbert and Sullivan records, as he likes to do, and looking out on the gathering darkness, he remembers other, earlier days when living was a greater adventure, and every morning brought with it a new and welcome challenge. And sometimes, in remembering those days, they would seem to be very near. He could recall so clearly the sharp strange smell of benzine at the garage in Longwall Street, the tight feeling of nervousness in his stomach when he had agreed to rent the old Military College at Cowley…measuring Mr Pilcher for his cycle… Bill Anstey in the loft, handing down newly painted wheels on the end of his rope…breaking down in the first Morris-Oxford with Gordon Stewart… As one grows older, things that happened long ago always seem much nearer than those that have taken place within the last few weeks or months…

Shortly after the war, Sir Miles Thomas left him. They had worked together since 1923. Some time before he resigned, Sir Miles had been appointed a director of the Government's Colonial Development Corporation, but he believed that this need not mean his resigning from the Vice-Chairmanship of the Nuffield Organization. However, events were to prove him wrong, and it became impossible for him to devote all his time to the Organization. He foresaw a tremendous job in Empire development, and in the following January went to South Africa to form a commission to co-ordinate development in Southern Rhodesia. Later, he resigned from this and joined the British Overseas Airways Corporation, of which he became Chairman shortly afterwards.

Lord Nuffield feels the gradual thinning out of his contemporaries, and the loss of those who were with him in the early days. Now, other younger men who were merely boys then, have taken their places.

"What friends I have, I value," he says, weighing his words. "The intimate side of a man's life, I find, tends more towards loneliness as his achievements increase. It is, I think, more a loneliness of spirit. Depending so much on himself, he develops a tendency to shut people out from his inmost thoughts..."

This is the loneliness that Lord Nuffield knows. "You just can't imagine what it is like to be able to have anything you want," he says, "to look in shop windows and be able to buy anything you see."

After the war he was still as active as ever in his factories, respected, venerated, even feared. He had become more accustomed to being a public figure, but still he was never quite at ease in ceremonies, and he disliked all personal publicity.

The sharpness of his comments had not mellowed with the passing years. They were still as pungent and as forthright as ever.

Somebody, about this time, asked him for his recipe for success, as though it were something he kept written down on a piece of paper in his wallet.

Nuffield took his cigarette holder out of his mouth, and looked intently at his questioner.

"Damned hard work," he said briefly, and, of course, he was right, although the word "luck" is sometimes mentioned, too, in this connection – a word that irritates him intensely.

"Nothing annoys me so much as hearing people say I've been lucky. I don't believe in luck – unless you really mean opportunity. And then you must still have foresight and courage to seize it."

What about success, then?

"My opinion," says Lord Nuffield, "is that no man can expect to have a very easy life and a very successful one. If he wants to lead a happy-go-lucky existence, he must forfeit success, and in so doing he may get more so-called enjoyment and rude health. But he must not at the end of his career claim that he has had no luck. He can't have his cake and eat it."

Could he repeat his success again, if he were starting now?

"It may still be possible to build a *business*," Lord Nuffield allows, "but I would say it is impossible to expand it as we were able to do in the past, with taxation and restrictions as high as they are at present, and with so many people doing as little work as possible for the highest pay, in the shortest time, and then grumbling at the high cost of living..."

He has never been able to reconcile himself to the philosophy of less work for more money that has gained so much ground since the war. He has always worked hard himself, and enjoyed the work; he cannot understand why so many of his countrymen have so little ambition. But then, as he admits, a bit wryly, "I look upon life essentially as a practical man. I have not much use for philosophies. Theories of any sort leave me cold. What knowledge, beliefs and feelings I have are the result of experience. I distrust any other sort."

Lord Nuffield is still unpredictable and spontaneous. In 1951, for instance, on his seventy-fourth birthday, which coincided with the production of the two millionth car to come off the assembly line at Cowley, about three hundred colleagues in the motor industry gave a dinner in his honour at a London hotel. As they stood and toasted this slight, white-haired man, four waiters wheeled in a trolley on which was a gigantic cake weighing ninety pounds.

Someone handed him a knife, and while six hundred hands clapped together enthusiastically, he got up to cut this birthday cake. He stood for a moment, the glittering blade in one hand, smiling round at the guests. Many of them were his personal friends: they had known each other since the days of the first Morris-Oxford, and before. Suddenly, he threw the knife down on the table in front of him.

"I won't cut it," he exclaimed, grinning. "If you eat it, it will make you all sick. Let's send it to the Great Ormond Street Hospital instead. Let the children enjoy it."

There followed a moment of stunned silence, eyebrows raising as head turned to head. Then, slowly at first, and louder and louder, the clapping began again. Only Lord Nuffield could have done this, but then only he would ever think of such a thing...

"What shall we do with this car, the two millionth Morris?" he was asked later, at the factory.

"Give it to the National Institute for the Blind," he replied at once.

There remained one master-stroke to make, the amalgamation of the Austin and Nuffield concerns; the biggest deal ever to be made in the British motor industry.

Under the agreement, a holding company, with an authorized capital of £5,000,000, would control the two organizations, with Viscount Nuffield as Chairman and his former colleague Leonard Lord – now Sir Leonard Lord – the Managing Director of Austin's, as deputy Chairman and Managing Director of the new company.

The total assets of the two firms amounted to £57,000,000, Austin's being £22,000,000 and the Nuffield group, £35,000,000. The issued capital of the Austin Motor Company was £4,669,672, and that of Morris Motors, £5,650,000.

It was stressed that none of the vehicles in the new British Motor Corporation would lose their identity or their individuality; all would continue to be sold under their own names.

Although the merger had been kept a secret, such an alliance might not have been unexpected, because of the need to make a stand against the fierce competition from America, where three big companies controlled most of the motor industry; and from

Germany, where the Volkswagen was being produced and exported in increasing numbers. Sharper competition was also to be expected from both the Ford Company at Dagenham and Vauxhall Motors at Luton, which were planning developments to the tune of millions. This merger seemed the best way to meet it.

Before this move, when friends used to suggest to Lord Nuffield that he retire, he would shake his head vigorously at such talk.

"I work hard still for two reasons," he would point out patiently. "One, because no businessman who has built up an organization employing tens of thousands of wage-earners can suddenly drop the reins without a thought for the vast family associated with his labours; two, because work is still the natural mission of every real man."

After the merger between the two great companies had been successfully completed, he surprised everyone by suggesting himself that he retire, and he announced his intention at the first annual meeting of the new Corporation, held in Oxford one year after its foundation, in December of 1952.

"In October I reached my seventy-fifth birthday, and I have now been at the helm for nearly sixty years," he said, and paused. Then he continued slowly, measuring out each word with care, looking at the faces round him.

"Although I am glad to say my health is good for my age, this seems an appropriate time for me to hand over my business responsibilities to a younger generation... I have therefore decided to resign from my directorship...but I have agreed to undertake the office of Honorary President of the British Motor Corporation.

"Thus my resignation from the boards will not mean a complete abandonment of the work which has always been the main interest of my life, and my advice and experience will always be available..."

Such a man can never really retire. He is still at his office in Cowley on most mornings, as he has always been, save when he visits his factory in Australia, as he has taken to doing in the winter months. He drives himself in his 1938 Wolseley Eight, or in the 25 hp coupé of the same year and make which the workpeople of Wolseley's presented to him some time before the war. There is a third car, also a Wolseley, known to his friends as "the limousine", and this he usually travels in when he is being driven on official business.

All the cars are black. "I'll have any colour you like, so long as it's black," he grins, modifying Henry Ford's famous phrase. He refuses to have a new car, although it has often been suggested that he should.

"These'll see me out," he says with finality. "They're good enough for me, and anyway, I like them..."

The best appreciation of Lord Nuffield was made not in English, but in Latin; and not recently, but more than twenty years ago, in 1931, when Oxford University conferred on him the honorary degree of Doctor of Civil Law.

In this ceremony, it is usual for the Public Orator of the University to make a short speech in Latin, in which he traces the main events of the life of the man about to receive the degree. In this case, the Orator recalled how Morris had caused the old horse-drawn trams to be superseded by buses, and he went on to describe how the name of this "*fortunae suae faber*" – "the maker of his own fortunes" – had been spread abroad by the Oxfords and the Cowleys, and he pointed out how "*inter miracula ipse praecipuum miraculum, inter moventia primum mobile*" – "among miracles, he is himself the chief miracle; among movement, he is the source of motion."

It was the perfect description.

Sir William Morris, as he then was, sitting in his new red robes, and uneasy in the public gaze, must have realized how apt the words were. The old alchemists and philosophers knew that the worlds around us – the stars, the planets, the other earths – were always in motion across the sky, and they sought the inspiration of this perpetual movement. The prime mover, they held, was the source of all motion; and so far as Cowley was concerned, this was certainly true of W R M. He was the mainspring of that enterprise, which, in unwinding, set great wheels to turn and so kept the whole machine alive.

He started it, and he kept it running; always remember that. There were others to help him, but he was the first; his was the conception of the idea, and he made it grow.

Yet, for his part, he is a man of strange contrasts: a life-long believer in the virtues of free, unfettered enterprise – and yet one who has given away nearly all his wealth to help his less fortunate fellows; a millionaire who is not interested in money; a man who loves children but who has none to carry on his name.

"Any fool can invent anything, as any fool can wait to buy the invention when it is thoroughly perfected," said Rudyard Kipling, a keen motorist himself, in 1904. "But the men to revere, to admire, to write odes and erect statues to, are those Prometheuses and Ixions (maniacs, you used to call us) who chase the inchoate idea to fixity up and down the King's Highway, with their red right shoulders to the wheel."

William Richard Morris stands at the head of this great and glorious company of motoring pioneers, a man who harnessed power to turning wheels – and rode those wheels to fortune.

JAMES LEASOR

BOARDING PARTY

Filmed as *The Sea Wolves,* this is the story of the undercover exploits of a territorial unit. The Germans had a secret transmitter on one of their ships in the neutral harbour of Goa. Its purpose was to guide the U-boats against Allied shipping in the Indian Ocean. There seemed no way for the British to infringe Goa's Portuguese neutrality by force. But the transmitter had to be silenced. Then it was remembered that 1,400 miles away in Calcutta was a source of possible help. A group of civilian bankers, merchants and solicitors were the remains of an old territorial unit called 'The Calcutta Light Horse'. With a foreword by Earl Mountbatten of Burma.

'One of the most decisive actions in World War II was fought by fourteen out-of-condition middle-aged men sailing in a steam barge…' – *Daily Mirror*

'A gem of World War II history' – *New York Times Book Review*

'If ever there was a ready-made film script…here it is' – *Oxford Mail*

JAMES LEASOR

GREEN BEACH

In 1942 radar expert Jack Nissenthall volunteered for a suicidal mission to join a combat team who were making a surprise landing at Dieppe in occupied France. His assignment was to penetrate a German radar station on a cliff above 'Green Beach'. Because Nissenthall knew the secrets of British and US radar technology, he was awarded a personal bodyguard of sharp-shooters. Their orders were to protect him, but in the event of possible capture to kill him. His choice was to succeed or die. The story of what happened to him and his bodyguards in nine hours under fire is one of World War II's most terrifying true stories of personal heroism.

'*Green Beach* has blown the lid off one of the Second World War's best-kept secrets' – *Daily Express*

'If I had been aware of the orders given to the escort to shoot him rather than let him be captured, I would have cancelled them immediately' – *Lord Mountbatten*

'*Green Beach* is a vivid, moving and at times nerve-racking reconstruction of an act of outstanding but horrific heroism' – *Sunday Express*

JAMES LEASOR

THE MARINE FROM MANDALAY

This is the true story of a Royal Marine wounded by shrapnel in Mandalay who undergoes a long solitary march to flee the Japanese and finds his way back through India to Britain. On his way he has many encounters and adventures and helps British and Indian refugees.

THE MILLIONTH CHANCE

The R101 airship was thought to be the model for the future, an amazing design that was 'as safe as houses…except for the millionth chance'. On the night of 4 October 1930 that chance in a million came up, however. James Leasor brilliantly reconstructs the conception and crash of this huge ship of the air with compassion for the forty-seven dead – and only six survivors.

'The sense of fatality grows with every page… Gripping'
– *Evening Standard*

JAMES LEASOR

THE ONE THAT GOT AWAY

Franz von Werra was a Luftwaffe pilot shot down in the Battle of Britain. *The One that Got Away* tells the full and exciting story of his two daring escapes in England and his third and successful escape: a leap from the window of a prisoners' train in Canada. Enduring snow and frostbite, he crossed into the then neutral United States. This book is based on von Werra's own dictated account of his adventures and makes for a compelling read.

THE PLAGUE AND THE FIRE

This dramatic story chronicles the horror and human suffering of two terrible years in London's history. 1665 brought the plague and cries of 'Bring Out Your Dead' echoed through the city. A year later, the already decimated capital was reduced to ashes in four days by the fire that began in Pudding Lane. James Leasor weaves in the first-hand accounts of Daniel Defoe and Samuel Pepys, among others.

'An engrossing and vivid impression of those terrible days' – *Evening Standard*

'Absorbing…an excellent account of the two most fantastic years in London's history' – *Sunday Express*

34857166R00083

Printed in Great Britain
by Amazon